P9-DWR-491

TINKERNUT

PRESENTS

UPCYCLED TECHNOLOGY

Published by Mango Publishing Group, a division of Mango Media Inc.

Cover and Layout Design: Elina Diaz

For permission requests, please contact the publisher at:

Mango Publishing Group
2850 Douglas Road, 2nd Floor
Coral Gables, FL 33134 USA
info@mango.bz

For special orders, quantity sales, course adoptions and corporate sales, please email the publisher at sales@mango.bz. For trade and wholesale sales, please contact Ingram Publisher Services at customer.service@ingramcontent.com or +1.800.509.4887.

Tinkernut Presents: Upcycled Technology: Clever Projects You Can Do with Your Discarded Tech

Library of Congress Cataloging

ISBN: (print) 978-1-63353-909-9 (ebook) 978-1-63353-910-5

Library of Congress Control Number: 2019932088

BISAC category code: SCI028000—SCIENCE / Experiments & Projects

Printed in the United States of America

TINKERNUT
PRESENTS

Clever Projects You Can Do with Your Discarded Tech

CORAL GABLES

TABLE OF CONTENTS

dissentiet vim.

Invidunt neglegentur cu per, labores gubergren voluptaria id duo. An ahsre maiestatis consequuntur pri. Homero munere nominavi eum no, an sit idque doctus tincidunt. Saepe legimus euripidis cum ut. Dicit tation scripserit vim no, illum commune scriptorem nec in. No vis dicam partem.

eu. I
ri, ea
petua periculа
Usu ut
endo
n hendrerit

a voluptaria
Homero munere
. Saepe
rit vim no,
partem.

INTRODUCTION

▶ Difficulty

WHY UPCYCLE OLD TECHNOLOGY?

If you are not familiar with the concept of "upcycling," I like to define it as the creative act of turning junk into something marvelous. There's something gratifying about taking what's been discarded and giving it a new soul. While upcycling is most commonly associated with clothing or furniture, we will be focusing on upcycling old technology. So does that mean we will be turning old keyboard keys into new earrings or old CDs into coasters? Not so much. This book will focus more on repurposing and reusing the technical components themselves, such as turning an old flip phone into a smartwatch, or an old laptop into a projector. Why would anyone want to do that? What are the benefits? Let's take a look!

Erasing E-waste

According to the EPA, only 12.5 percent of e-waste is recycled.[1] What happens to e-waste that doesn't get recycled? It gets sent to landfills. A report conducted by the United Nations Environmental Program[2] estimated that fifty million tons of computers, smartphones, and other electronic waste were sent to the dump annually. This can have devastating effects on the environment not only from e-waste, but also through increasing the mining of precious metals that are used to create the newer devices.

---

1 https://earth911.com/eco-tech/20-e-waste-facts/
2 http://web.unep.org/ourplanet/september-2015/unep-publications/waste-crime-waste-risks-gaps-meeting-global-waste-challenge-rapid

Saving Some Money

There are always cost saving benefits when you decide to upcycle as opposed buying something brand new. As a tinkerer, however, it's hard for me to see all those millions of tons of transistors, resistors, processors, LCDs, LEDs, and other useful bits and bytes that just go to waste. There is so much value and potential just waiting for the right creative spirit to scoop it up and transform it into something incredible! Lesser known pro tip: most electronics use precious metals in their components. It's estimated that each year Americans throw away sixty million dollars in gold and silver. Just think about learning how to extract that!

Kindling That Creative Spark!

I firmly believe that the best way to learn is to embrace creativity and curiosity. A wise man once said, "Creativity is intelligence having fun." Upcycling technology is a great way to tap into that creativity while having fun and learning more about the technological world around you. Taking a peak behind the casing helps demystify technology while making it less scary and confusing yet more intriguing and approachable. Seeing the creative potential in old technology can inspire you to create anything; you're limited only by your imagination! Have I piqued your interest yet?

Let's jump right in!

dissentiet vim.

Invidunt neglegentur cu per, laoreet gubergren voluptaria id duo. An choro maiestatis consequuntur pri. Homero munere nominavi eum no, an sit idque doctus tincidunt. Saepe legimus euripidis cum ut. Dicit tation scripserit vim no, illum commune scriptorem nec in. No vis dicam partem.

eu. I
ri, ea
petua pericula
Jsu ut
endo
a hendrerit

a voluptaria
omero munere
Saepe
it vim no,
artem.

PREPARE TO REPAIR

SAFETY FIRST!

Manuals are for putting things together, not taking things apart, right? We're about to go beyond the instruction booklets to see where the world of technological upcycling can take us! But even seasoned upcyclers need to be cautious when dealing with electronics. Messing around with old devices doesn't come without its risks. Once we alter how something works and plug it into a power source, improperly connected components can be harmful or deadly. So to help alleviate becoming human toast, please be sure to tap into your common (and uncommon) senses when creating your own devices. Along with that, here are a few tips to help you stay safe.

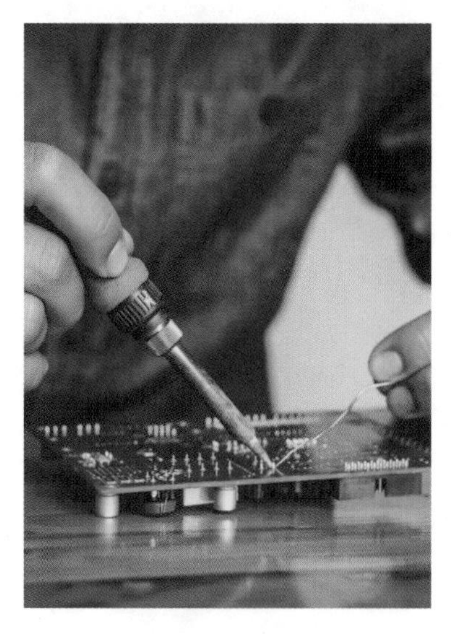

Red Wire or Green Wire?

If you've never cracked open an electronic device before, staring at all those wires, circuits, and components can be a little bit daunting. To help demystify the unfamiliar land of electronics, it would be beneficial to know at least some of the components you're looking at so that you can navigate this new territory easily. Not only is this knowledge recommended for safety reasons (since for example, some components can store electrical charges), but it also aids in scavenging for useful parts and components.

Where's That Fire Extinguisher?

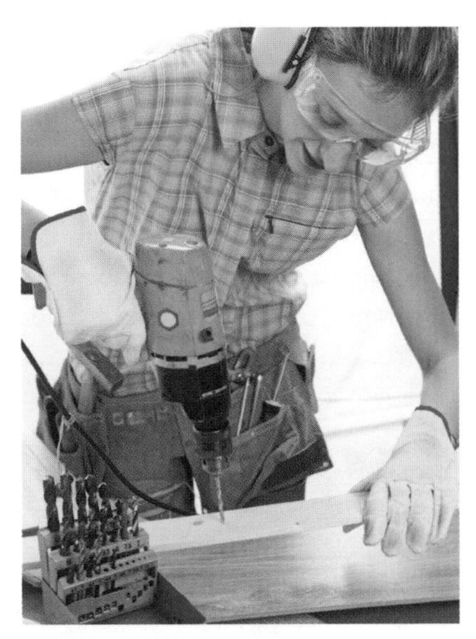

Aside from what you have decided to work on, where you decide to work on it is just as important. Make sure that your workspace has plenty of ventilation and is safe and dry. Try not to have too much clutter or debris that can get in the way. First-aid kits and fire extinguishers are always a good thing to have. Also, wearing proper safety equipment such as goggles, antistatic gloves, face masks, and so on is a good idea. Hey, if we're channeling our inner mad scientists, we might as well look the part, right?

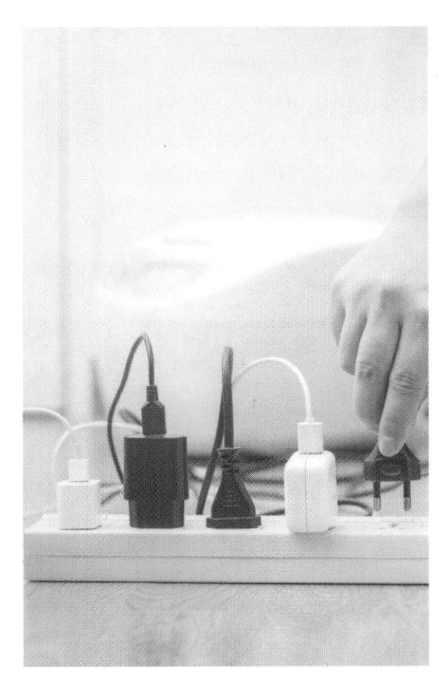

Electricity Is DANGEROUS

Although you probably already know this, it's important to point out that electricity can be very harmful or deadly. *Always* disconnect power before working on a device. Capacitors can store electricity long after a unit has been powered off, so it's important to learn how to properly discharge them before handling them. Use surge protectors and *never* plug anything directly into a wall outlet. And if you smell something burning, disconnect power *immediately*.

WHERE DO I LOOK?

When getting started with upcycling old electronics, the first thing you need is "stuff." Since computers, tablets, and smartphones are constantly being upgraded, they creates a surplus of old tech just waiting to be saved from a death by dumpster and given a new life. That being the case, it shouldn't be too hard to find what you need for free or super cheap. The key is knowing where to look, so I've put together some of the most common places to find old tech.

Home

Most of you reading this book probably have a junk drawer, closet, or attic full of hidden tech gems primed for hacking and upcycling (I had a Radio Shack Tandy 1000). The obvious benefit of starting from home is that it's free. And since you own it, you are probably already somewhat familiar with it. Another bonus is that using it only requires permission from yourself.

Dumpster Diving

It's surprising how much e-waste actually makes it into the trash. But we can turn that tragedy into our gain, not to mention, it's free! Dumpster diving is the act of rummaging through dumpsters to find hidden treasures. In this case,

those treasures are old or broken electronics. College campuses (especially Ivy League colleges) are great places to find useful stuff tossed out by a student who doesn't have a need for it anymore. If those tech treasures end up making their way to the landfill, not to worry! Your local dump itself can be a great place to look for goods.

Yard Sales

Love a good yard sale? If dumpster diving isn't your thing, then yard sales are a great way to score decent and unique electronics for cheap. A lot of sellers don't know the actual value of what's being sold and ultimately just want to get rid of it. The downside is that yard sales can be infrequent and hard to find. Also, the items generally aren't free, but sometimes it may be worth spending a few bucks for the perfect item.

Online

Let's be real, visiting yard sales and dumpster diving can require a lot of effort. If that's not for you, then acquiring devices online is probably the way to go. You may not be able to find old broken devices on Amazon, but sites like Craigslist or eBay (in the US) may have the perfect item for your project!

WHAT DO I LOOK FOR?

Now that you know where to look, how do you know what to look for? Sorting through dozens of electronic items to find what you need can be daunting if everything just looks like a pile of odd-looking bits and pieces. Some devices are richer in reusable components than others, so it's helpful to know which devices are guaranteed to be a good score. One key piece of advice I can offer is that older devices tend to have components that are easier to reuse. Newer devices have smaller and more integrated components that are more difficult to extract. With that in mind, here are a few devices that are worth considering if you come across them.

Printers, Scanners, Faxes

Old printers, scanners, and fax machines are great finds because they contain reusable motors and generally a sliding mechanism. Most of them will also have buttons and maybe an LCD screen or two that can be salvaged. Scanners also have really cool light bars that can be used for...something.

VCR and DVD Players

With how quickly media formats change, those poor VCR and DVD players went from being heroes to zeroes in a little more than a decade. But lurking inside those old video titans are motors, infrared LEDs, LEDs,

buttons, and some cool female RCA jack adapters. There are tons of scavengable items in these devices!

Routers and Modems

Remember the good ol' days of dial-up? Even though these nostalgic devices aren't as useful themselves, they are chock-full of easily extractable components that are just begging to be reused. Expect to find tons of resistors, capacitors, diodes, and LEDs.

WHAT TOOLS DO I NEED?

When it comes to taking apart, modding, or scavenging old electronics, unless you are Edward Scissorhands, you will probably need more than just your nimble fingers to get the work done. Manufacturers all have unique ways of assembling products, which means that the ease of taking an item apart varies from device to device. For some reason, screws have become something to avoid in newer devices, and the trend has been to glue them together. Sadly, that makes newer smartphones and tablets more difficult to disassemble. In general, however, these tools should get you through most upcycling projects.

Rotary Tool

A rotary tool is a small handheld tool with a tip that rotates very fast. The rotating tip can be interchanged depending on the task. It is often used for cutting, sanding, grinding, and polishing but can be used for a lot more tasks than those.

Soldering Iron

A soldering iron is a tool that can heat up to high temperatures in order to melt a form of metal alloy wire called "solder." Since solder is conductive, melting it onto a circuit board can join two

circuit components together. For home electronics enthusiasts, this is a necessary tool for creating homemade circuits.

Aside from joining circuits by melting solder, soldering irons can also assist in removing circuit components by melting the solder around that component.

Wire Cutters/Strippers

Wire strippers are handheld tools that are primarily used to remove the plastic sheathing from wires. This exposes and allows access to the metal wire underneath the plastic. Used in conjunction with a soldering iron, these exposed wires can be used to connect circuit components together.

Multimeter

Multimeters are instruments that can measure and display voltage, electrical current, and sometimes resistance (depending on the multimeter). Each has a + probe and a - probe that are used to connect to the positive and negative leads on a power source. The multimeter will then tell you the value of the current, voltage, and/or resistance.

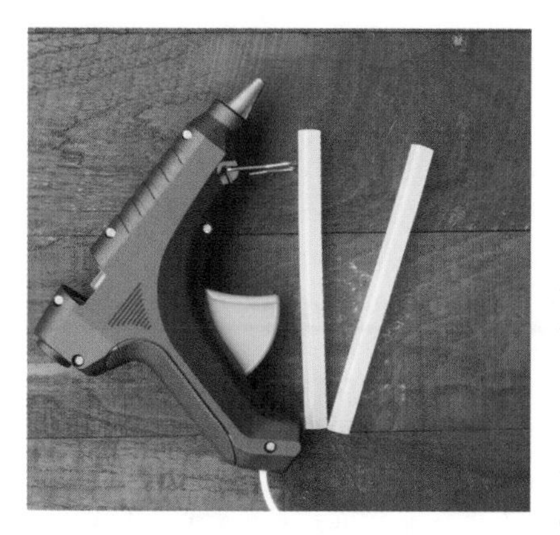

Hot Glue

"Hot Glue" is a term used to describe solid glue sticks that require melting in order to be used. Hot glue "guns" are used to melt the glue sticks. They have a tapered nozzle to make it easy to apply the glue where needed. Once the melted glue is applied, it takes a couple of minutes to dry. Dried hot glue forms a strong, solid bond. That makes it useful for repairing and joining different materials.

Arduino

An Arduino is a popular microcontroller that is easy to program and implement. Microcontrollers are small, simple computers commonly used to perform single, specific tasks. In the case of the Arduino, it can be programmed to perform a task, and then it will continue to loop through that task until it is powered down or the program is changed. This makes it useful for controlling motors, lights, buzzers, sensors, and other simple electronic components.

Raspberry Pi

Raspberry Pi is a brand of small form factor computers. These computers range from the size of a credit card to the size of a pack of chewing gum. They operate similarly to larger computers in that they have a desktop interface, audio and video output, and USB and Ethernet ports. The Raspberry Pi also features Input/Output pins that can be programmed to control different electronic components (similar to an Arduino). The largest difference between the Raspberry Pi and the Arduino is that the Raspberry Pi can perform multiple processes at the same time instead of one at a time.

dissentiet vim.

Invidunt neglegentur cu per, labores gubergren voluptaria id duo. An choro maiestatis consequuntur pri. Homero munere nominavi eum no, an sit idque doctus tincidunt. Saepe legimus euripidis cum ut. Dicit tation scripserit vim no, illum commune scriptorem nec in. No vis dicam partem.

eu. I
ri, ea
petua pericula
Usu ut
endo
an hendrerit

n voluptaria
Homero munere
. Saepe
rit vim no,
partem.

COMPUTERS & PERIPHERALS

▶ Difficulty

Do you remember your first computer? Perhaps it was an IBM running DOS, or were you more of a Commodore 64 person? Maybe you were one of those Apple II renegades? Mine was a Radio Shack Tandy 1000 complete with a 7.16MHz processor running MS-DOS and a 3¼ inch floppy drive! That classic has been obsolete for a few decades now. It's crazy to imagine how quickly computers and their peripherals become outdated. Remember the good ol' days of floppy disks, zip disks, CRT monitors, data cassettes, USB webcams, and [insert favorite antiquated computer reference here]?

Aside from playing old computer games, there isn't much utility in really old computers in their original state. However, some computers and peripherals can be repurposed and refurbished into something different and more useful. It all depends on what you have and how creative you are with it. We'll take a look at three projects ranging from beginner to more advanced, so no matter what the level of your electronics expertise is, you should be able to create something cool!

PROJECT 1

OLD WEBCAM TO BACKUP CAMERA

Beginner

Synopsis: Back before smart devices, computer cameras and webcams were external devices that you would have to plug in through USB. Since most computers nowadays have them built in, what do you do with your old external webcam?

Old Webcam to Backup Camera - Larger images can be found on page 102 of photo glossary

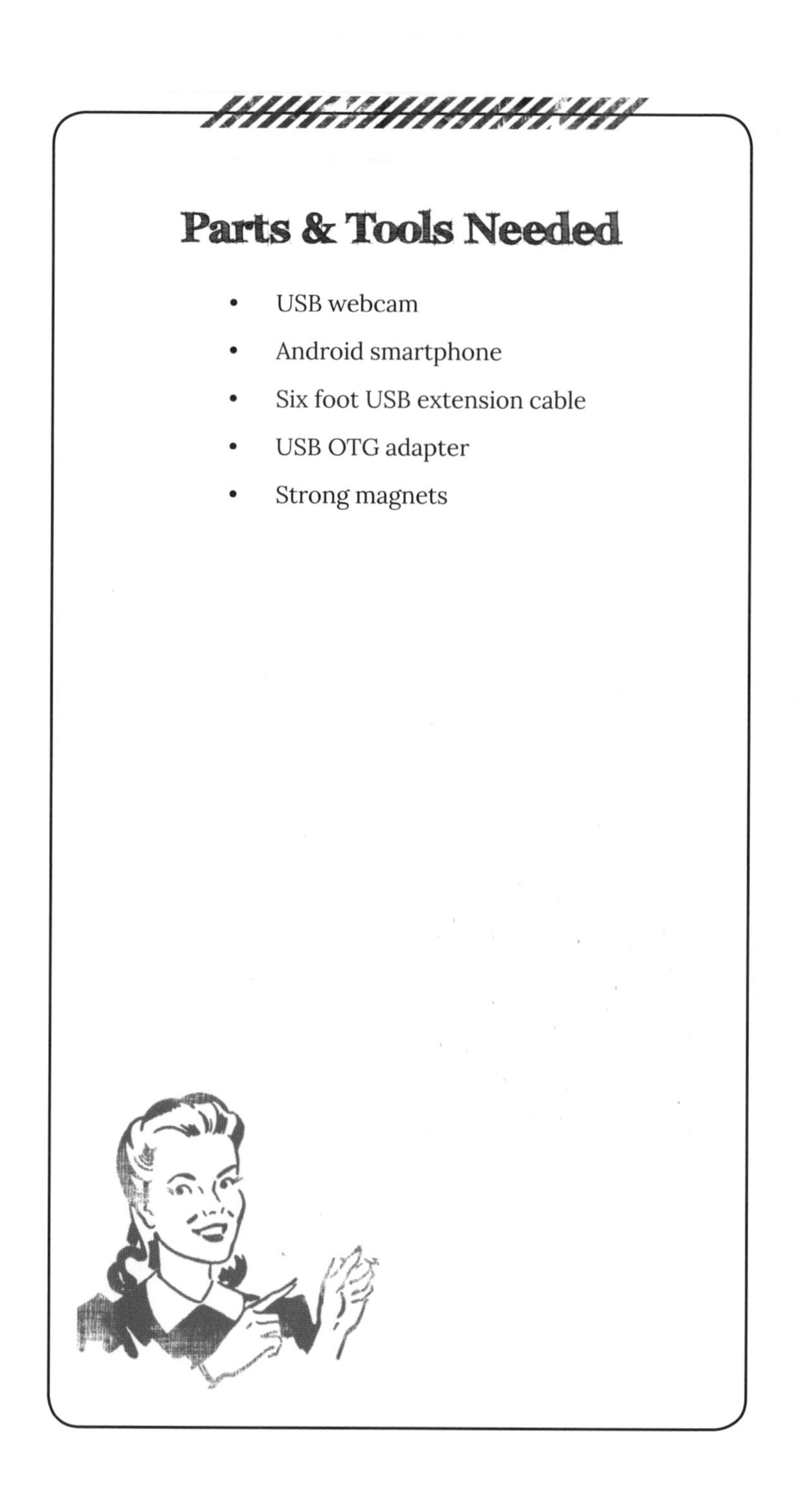

Parts & Tools Needed

- USB webcam
- Android smartphone
- Six foot USB extension cable
- USB OTG adapter
- Strong magnets

Step 1: Connecting the Webcam to a Smartphone

I have at least a couple old webcams that I used pretty frequently before cameras became integrated into almost every device you can think of nowadays. I wanted to see if it would be possible to use one of them as a backup camera for my car. Let me preface that with a huge **disclaimer** that this is not intended for practical use and should not be relied upon as you would depend on a commercial backup camera. The intent is only to use this in rare instances, like getting close to a trailer for connecting it or backing up close to a dock for loading. So please, please, please don't rely upon this as a sole backup camera for your vehicle.

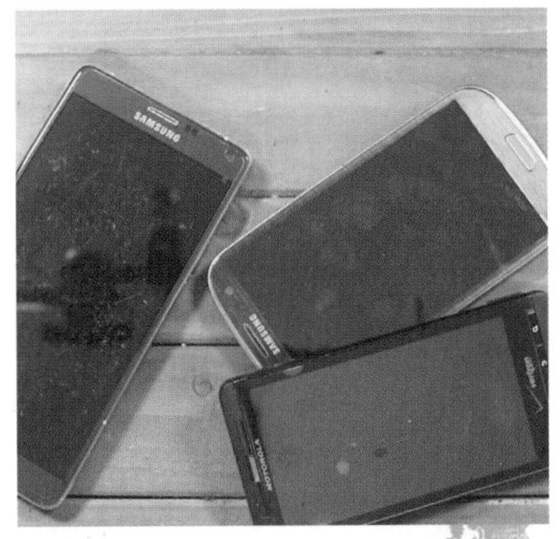

That said, the first thing that we will need to do is find some way to view the webcam from the driver's seat while the camera is mounted to the back of the car. Since I had several old Android smartphones lying around, I decided to use one of those as the viewing screen. Android devices made within the last five years have the ability to recognize USB devices when they are plugged in. This is known as "USB On the Go" or USB OTG. So we will need a USB OTG cable to connect the webcam to it. Just plug one end of the OTG cable into the webcam and the other end into the phone.

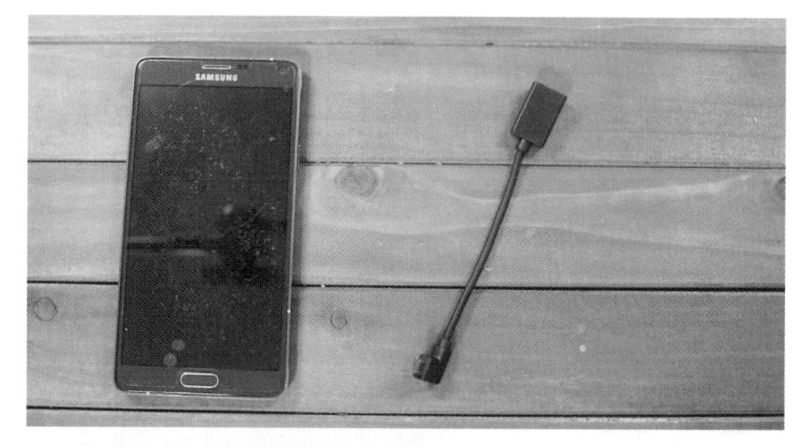

It's not enough to connect the webcam to the phone, we also need to have some type of software so that the phone will recognize the webcam and display video. Although they are in no way affiliated with this book, the Free USB Camera app by ShenYao[3] worked great for my needs. Once the webcam was plugged in, I just launched the app and the live video display from the webcam started right up!

3 https://play.google.com/store/apps/details?id=com.shenyaocn.android.usbcamera

Step 2: Installing the Webcam as a Backup Camera

The next trick is getting the webcam to stick to the back of the car while keeping it connected to the smartphone that will be at the front of the car. As for getting it to stick to the back of the car, I hot-glued a couple of strong rare-earth magnets to the base of the camera. After finding a good position on the back of the car, I can just put the camera on the car and the magnets will keep it stuck in place.

To connect the camera to the smartphone at the front of the car, all we need is a long USB extension cable. Those are pretty cheap to come by and vary greatly in length. The one I purchased is a six-foot extension cable. The cable ran into the trunk, through the side of the back seat, then along the floor under the front seat, where it ran up to the dashboard and connected to the camera. Even though you might think otherwise, the trunk was able to securely close and lock even with the USB cable running into it.

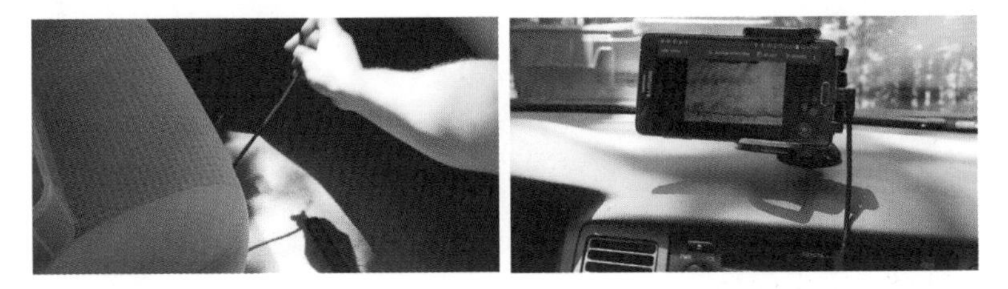

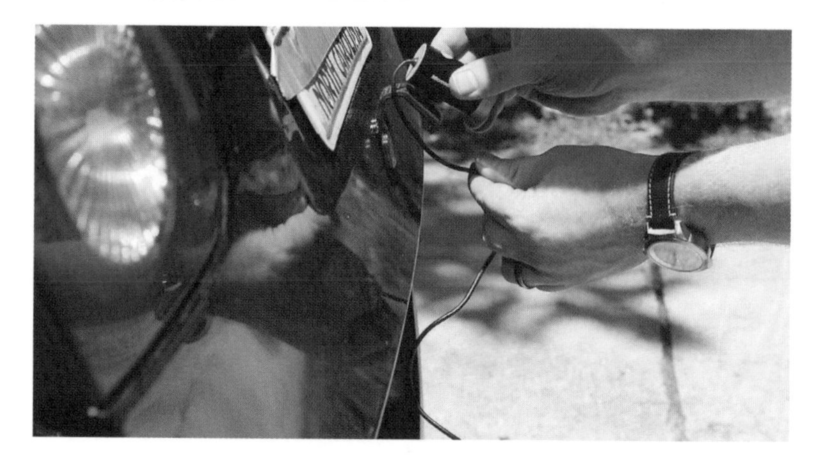

Step 3: Testing It Out

If everything is connected correctly, and the USB Camera app is open, you should see the camera image on your phone! You may need to adjust the positioning of the webcam so that it captures the area you need it to display. Now, in a safe location, you can back up and you should be able to see what's behind you! Again, let me stress not to use this as a commercial backup camera. However, it's a great little backup camera to use in a pinch, and it will probably fit in your glovebox!

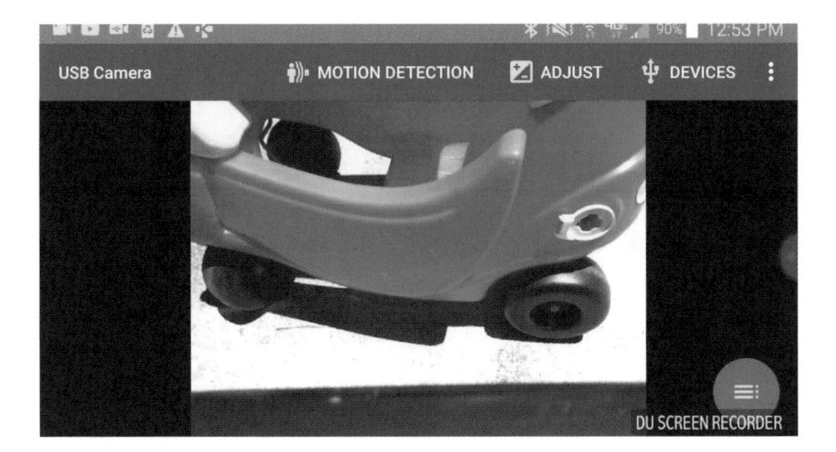
USB Camera
MOTION DETECTION
ADJUST
DEVICES
DU SCREEN RECORDER

PROJECT 2

TURNING AN OLD LAPTOP INTO A PROJECTOR

Intermediate

Synopsis: Got an old unused laptop lying around? With a few tools, you can turn that old laptop into a projector and watch movies on your wall!

Turning an old laptop into a projector - Larger images can be found on page 108 of photo glossary

Parts & Tools Needed

- Old unused laptop
- Overhead projector
- Screwdriver

Step 1: Finding a Laptop

As an IT professional, my family and friends ask me to fix their computer problems all the time. It's actually something I enjoy doing. In some instances, however, it's easier to buy a new computer instead of having the old one fixed. When that happens, I am quick to ask if I can have the old computer to repurpose. It's amazing how many "broken" laptops I've acquired and have been able to restore or repurpose.

For this project, I would only recommend using a laptop that has run its course, one that you don't mind if it gets ruined. The only requirement is that it has to have a functioning screen and be able to play video and audio files. The laptop I used for this project was one a friend gave me because the charging port was broken and it would no longer charge. With a screwdriver, soldering iron, and some hot glue, I was able to get it charging again, and it was ready to be revived!

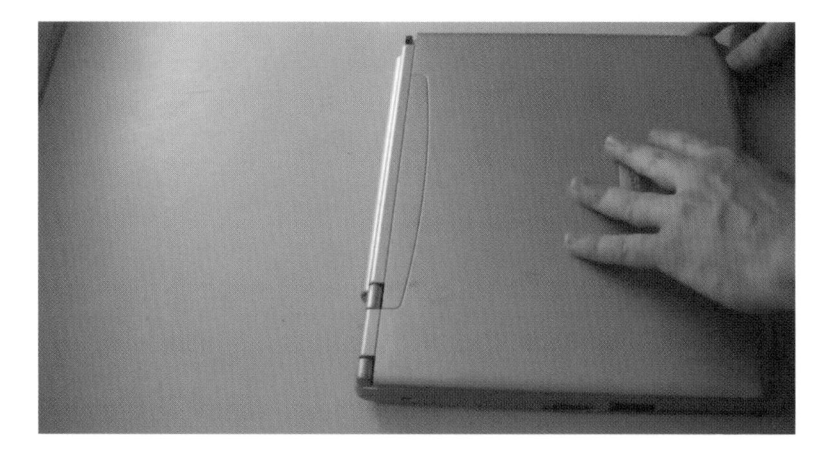

Step 2: Preparing the Screen

The trick for turning a laptop into a projector is all in the screen. That said, we need to extract the laptop screen and see what we have to work with. The screen is *very* fragile, so take these steps with caution. Most laptops have what's called a "bezel" around the screen, and generally that can easily be unscrewed and removed. If you are unsure how to go about doing that, you can probably find instructions online for your specific laptop by doing a Google search for your laptop model followed by the words "tear down." With the bezel removed, you should see how the laptop screen is attached to the laptop. Since every laptop is different, I can't really go into specifics on how to remove it, but generally it's basically removing a bunch of screws until you can carefully lift up the screen.

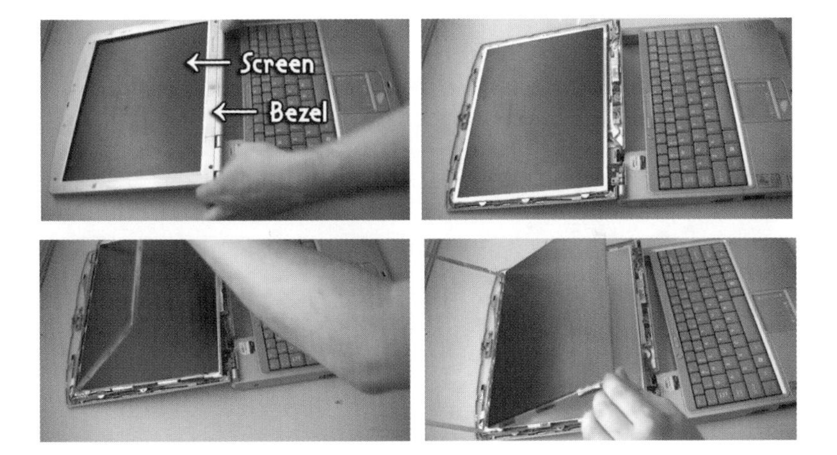

The screen should still be connected to the laptop with wires. These should remain connected and unbroken. You'll probably note that the screen actually consists of several different layers. The very back layer is generally a mirror and/or a backlight. The remaining layers are different filters and glass that produce the image. You'll see different wires, ribbons, and other circuitry around the screen that deliver power data to it. For my screen, the circuitry was all at the top of the screen on the back, so all I had to do is carefully swing it out so that it's not directly behind the screen. For some screens, the circuitry wraps around the top and side of the screen. If this is the case for you, this project may not work, as you won't easily be able to remove the circuitry from the back of the screen.

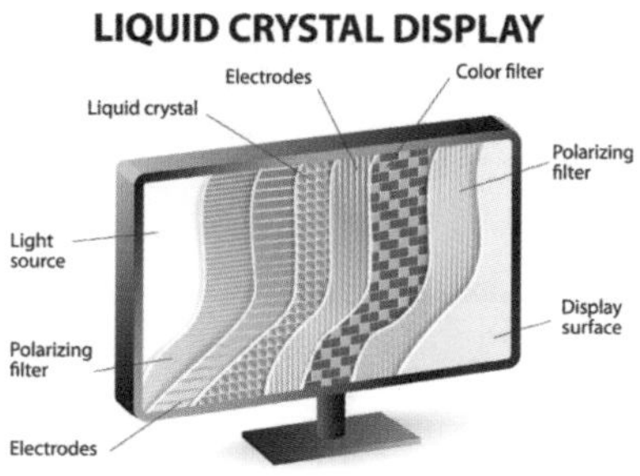

With the circuitry out of the way, the next thing I did was turn the computer back on. Normally, this is not something I would recommend doing, so again please be careful. Hopefully the screen still works and you can see images.

At this point, we can lift up the different layers of the screen to see which ones are required for it to display and image and which ones aren't necessary (such as the backlight). Basically, whenever you can lift up the screen and see your hand behind it while it's still displaying an image, then that's how you know what layers to keep.

Step 3: Projecting the Video

Now that we have the screen sorted out while retaining its ability to display videos, next we need a way to project the screen onto a wall. Back when I was in grade school, teachers had what were called "transparencies," which were basically transparent plastic sheets that they could write on using dry-erase markers. Then they would use overhead projectors to project those transparencies onto the wall. I didn't have an overhead projector, but I was able to purchase a working one off of Craigslist for twenty dollars.

All I have to do now is take the screen (wires and circuitry still connected) and lay it on the face of the overhead projector. Luckily the width of my screen was the same width as the overhead projector, so it was a perfect fit! Turn on the computer, and then turn on the projector. Shine the projector on the wall and adjust the focus until you can clearly see what's on your screen. Believe it or not, that's it! Now just plug in a nice set of speakers and fire up a nice movie on the "big" screen! One thing you might notice is that there may be large light leaks below the screen from where it doesn't quite cover the overhead projector surface. This can easily be fixed with some cardboard or construction paper to cover up those areas.

Light leak

PROJECT 3

CD-ROM DRIVE TO 3D PRINTER

Advanced

Synopsis: Ever wanted a simple 3D printer of your own just to see what the 3D printing craze is about? Instead of shelling out hundreds of dollars for one, you can build a basic one using an Arudino microcontroller and some CD/DVD-ROM drives.

Parts & Tools Needed

- 3 x Desktop computer CD-ROM or DVD-ROM drives
- 3 x Stepper motor drivers
- 1 x Arduino
- 1 x Desktop computer power supply
- 2 x Electrical box covers
- 1 x Generic 3D printing pen
- 1 x 22 Ohm resistor (value may vary depending on your 3D printing pen)
- 1 x NPN transistor
- Various nuts, bolts, and spacers
- Soldering equipment and wire

Step 1: The Basic Idea

I'm always surprised at how science fiction can inspire real life technology. In the original *Star Trek* and *Star Trek: The Next Generation* series, there is a device called a "replicator" that can materialize and reproduce many different objects nearly instantly.

Twenty years later, current 3D printers aren't that far from the technology imagined in the Star Trek replicators. In fact, there's even a line of 3D printers called "Replicators."

3D printers are an amazing technological innovation. The concept of 3D printers is a relatively simple one. You have a three-axis device (X axis, Y axis, and Z axis) that lays down a thin layer of melted plastic (known as filament) and continues to build up tiers of plastic layers until an object is formed. While the 3D printer market is still somewhat expensive, and while 3D printing devices are pretty complex, we can use the basic principles to make our own basic 3D printer using parts that can be scavenged from a desktop computer. Depending on what spare computer parts you have lying around, you can build this project for less than one hundred dollars.

Step 2: Disassembling the Drives

A 3D printer requires having a platform that can move on three axes. Typically referred to as the X axis, Y axis, and Z axis, all this really means is that the platform can move up and down (Z axis), to the front and back (Y axis), and to the left and right (X axis).

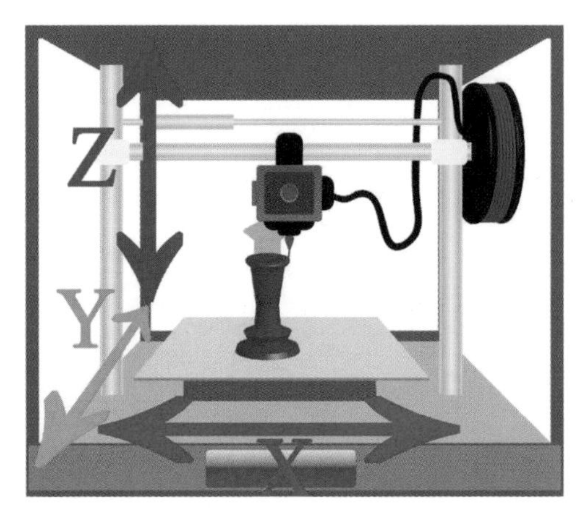

If you're old enough to remember CD or DVD-ROM drives (a.k.a. optical drives) on desktop computers, they were known for having a "tray" that would slide out whenever a button was pressed so that you could insert your optical disk. Then if you pressed the button again, the "tray" would slide back in.

The mechanism that makes this work is a little motor that moves the tray back and forth on a small track. This sliding motor tray mechanism is the perfect thing to use as a single axis for our 3D printer. And since there are three axes (X, Y, and Z), we will need three optical drives to match.

Taking apart the optical drive was actually a lot easier than it looked. The first thing to do is pull out the plastic drive "tray" and snap off the front panel by pulling the bottom of it forward slightly and then pushing it up. When that's done, flip the drive upside down and unscrew the bottom plate. Then it's just a matter of prying off the plastic front panel and metal casing.

CD-Rom Drive to 3D Printer - Larger images can be found on page 108 of photo glossary

With the outer casing off, you can see all the beautiful guts that make this thing tick: motors, lasers, LEDs, gears, all sorts of cool stuff that can be scavenged for use in other projects. For this project, we're interested in the metal mechanism with the spiral stepper motor and the plastic tray that slides back and forth on the track. The reason we want this specific part of the optical drive is because it offers a motor, track, and housing that can mechanically provide a smooth back and forth movement, which is ideal for a CNC axis. So you will need to disconnect any wires leading to the motor tray and separate it from the rest of the optical drive parts. You can remove the spindle motor (that spins the optical disk) from the tray if it is attached. You'll also want to remove the laser and any other glass parts, magnets, or stray pieces from the laser sled to make sure that it doesn't have anything that can hinder movement or mounting of other screws.

Now that the sliding motor tray mechanism has been extracted, analyze the stepper motor and make sure that it has four wires that lead to the stepper motor. (Note that if your drive does not have a four-wire stepper motor, then this project will not work.) We need the motor wires to be at least six inches long. If the current wires are not that long, then they need to be extended. I decided to desolder the old motor wires and solder on four new wires. I used different color wires for each motor so that I could tell them all apart.

Repeat this process for all three optical drives so that you have three bare motor trays; then you are ready for the next step.

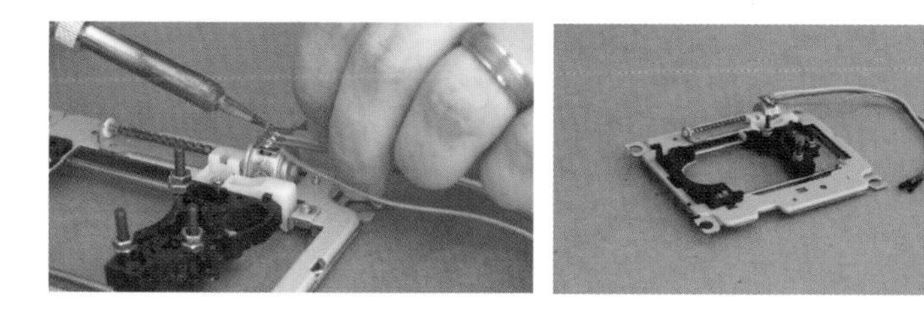

Step 3: Mounting the Motor Trays

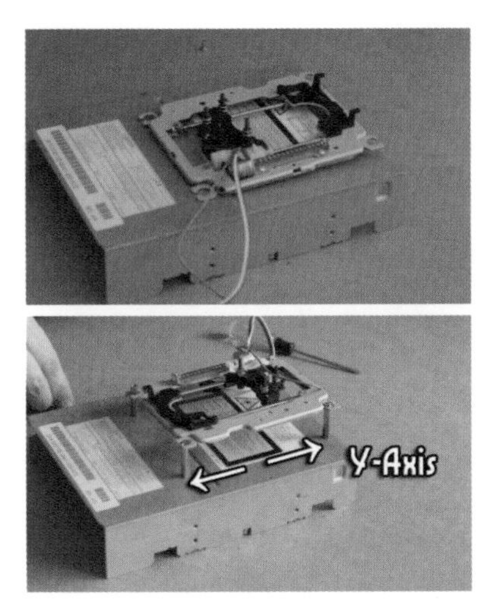

In order to achieve the X, Y, and Z axis movement, we will need to mount the sliding motor tray mechanisms onto some type of structure. Since we have three metal CD/DVD-ROM cases, I decided to use those as my mounting structure. Let's start with the Y axis. The Y axis will go back and forth, so take one of the motor trays and mount it parallel to the length of the casing close to one end.

Making sure it's aligned as straight as possible, use some screws to mount it while also making sure the tray will still slide unhampered. Then use bolts, bolt spacers, and nuts to mount the slide tray to the metal casing.

For the X axis, mount it perpendicular to the length of another optical drive case, again making it close to one end and aligning it as straight as possible. Then mount it using motherboard mounting screws as well. As for the Z axis, we will need to mount to the plastic tray sled of the X axis. In order to do this, I first mounted the Z axis to a salvaged metal plate (though you can use whatever you have available) using some bolts and spacers. Taking the entire Z axis plate, I attached it to the plastic tray on the X axis, again, using bolts and spacers.

Once you have all the motor trays mounted, the final step is to attach the X axis and the Z axis to the Y axis. You want to mount the X axis perpendicular to the Y axis (it will look like an "L" shape) and adjust them so that the Z axis is aligned over the Y axis. Scrub through each axis, checking carefully to make sure none of them are overshooting or running into each other. After you have the alignment set, screw everything together. I ended up using an L Bracket, but you may be fine just screwing one case directly into the other case. As a finishing touch, I added a flat metal plate (another piece of scavenged metal) to the Y axis so that the metal plate gave the Y axis a flat surface to print things on.

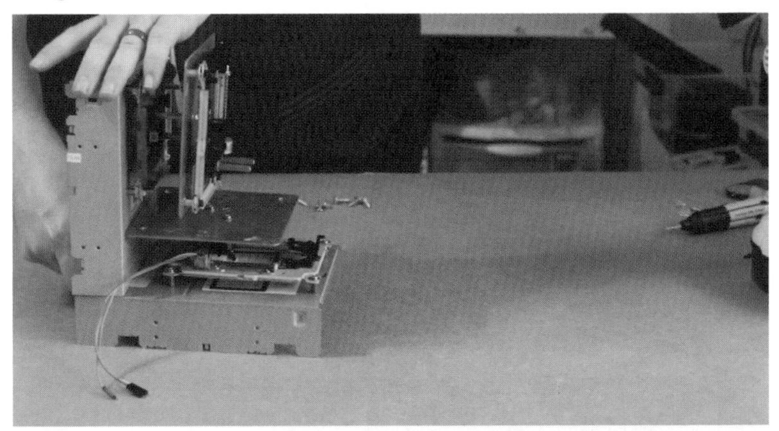

Step 4: Connecting the Electronics

At this point we have the equivalent of a car with no ignition and no power. We have our three motor trays, but we need something that can control them and make them move. An "Arduino" is an electronic microcontroller that is very simple to use even for beginners. It can be used to control lots of different electronic components, such as LEDs, LCDs, buttons, sensors, switches, and, in our case, stepper motors. While the Arduino has the ability to control the stepper motors, it doesn't have the capacity to provide enough energy to power them. To fix that problem, we will need what's known as a stepper motor "driver" for each motor. In the diagrams, you can see how to wire up each motor driver for each axis.

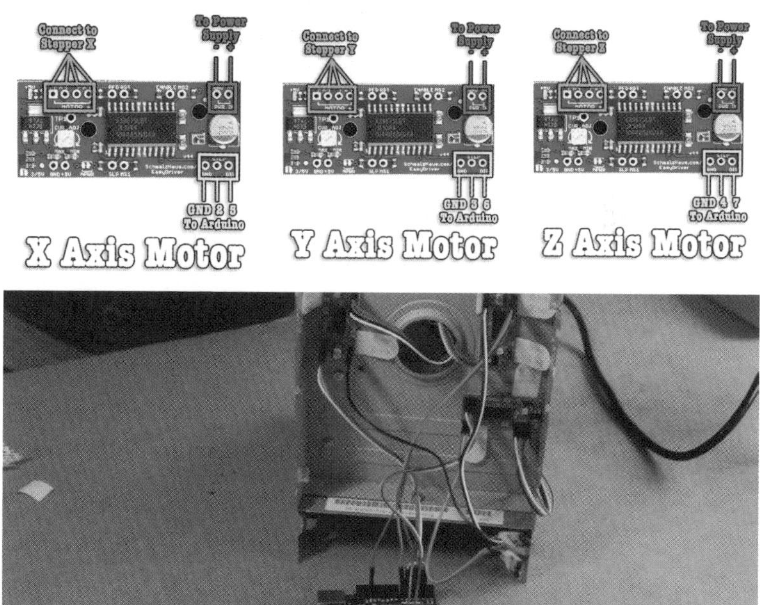

You may have noted that the motor drivers have two connections that are supposed to go to a "power supply." This is a power source that provides extra power. Since I'm mostly using parts from scavenged computers, my power source is going to be an old computer (ATX) power supply. A computer power supply can output several different specific types of voltage. Despite their usefulness, let me warn you that computer power supplies can be dangerous if you mess with them and don't know what you're doing. Please use caution when utilizing them.

20 PIN CONNECTOR

Pin		Pin	
(+3.3V)	1	11	(+3.3V)
(+3.3V)	2	12	(-12V)
(Ground)	3	13	(Ground)
(+5V)	4	14	(PS-ON)
(Ground)	5	15	(Ground)
(+5V)	6	16	(Ground)
(Ground)	7	17	(Ground)
(PG)	8	18	(-5V)
(+5VSB)	9	19	(+5V)
(+12V)	10	20	(+5V)

24 PIN CONNECTOR

Pin		Pin	
(+3.3V)	1	13	(+3.3V)
(+3.3V)	2	14	(-12V)
(Ground)	3	15	(Ground)
(+5V)	4	16	(PS-ON)
(Ground)	5	17	(Ground)
(+5V)	6	18	(Ground)
(Ground)	7	19	(Ground)
(PG)	8	20	(-5V)
(+5VSB)	9	21	(+5V)
(+12V)	10	22	(+5V)
(+12V)	11	23	(+5V)
(+3.3V)	12	24	(Ground)

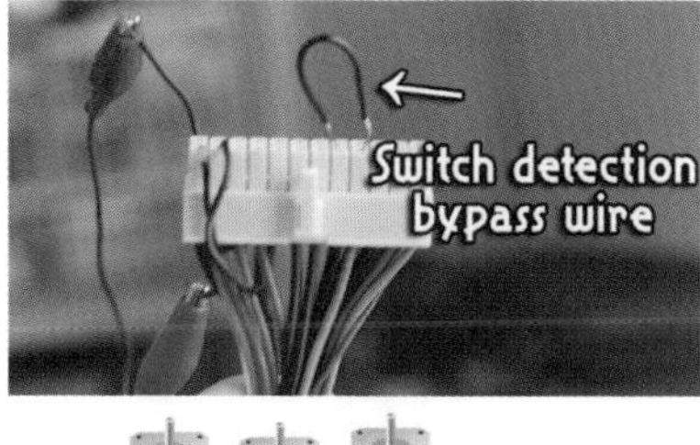

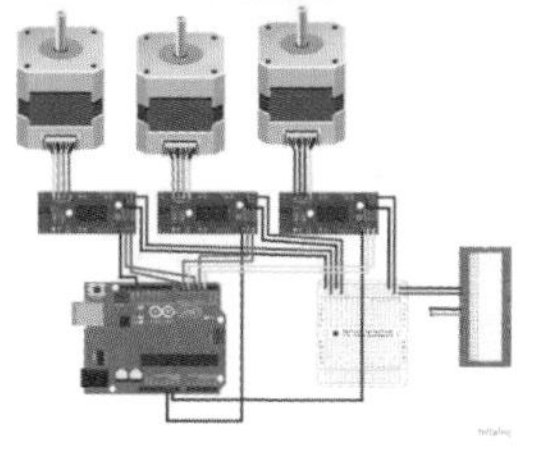

When being used inside of a computer, a computer power supply turns on and off whenever you press a power button. In order to use a computer power supply without a power button, we need to bypass its power switch detection. Most common ATX computer power supplies have either twenty or twenty-four pins. Depending on which version you have, you can use the guide to determine where the "PC-ON" and ground pins are located. Then connect a ground pin to the "PC-ON" pin using a scrap piece of wire. Once the switch detection is bypassed, you can also use this wiring diagram to determine where the ground and 5v pins are so that you can connect them to the motor drivers. This is what we will be using to power our motors.

Step 5: Hacking the 3D Pen

The essence of most hobby 3D printers is the ability to melt plastic filament and extrude it into layers. This requires a device known as a "hot end" that takes plastic filament, melts it, and outputs it onto a platform. There are a lot of commercial grade "hot ends" on the market, but they can be somewhat expensive, and they're kind of overkill given the homely nature of our build. Initially, I wanted to make my own "hot end," but it ended up being a difficult thing to fabricate at home. So what I did instead was purchase a cheap 3D printing pen to see if I could tweak it and make it work as the "hot end."

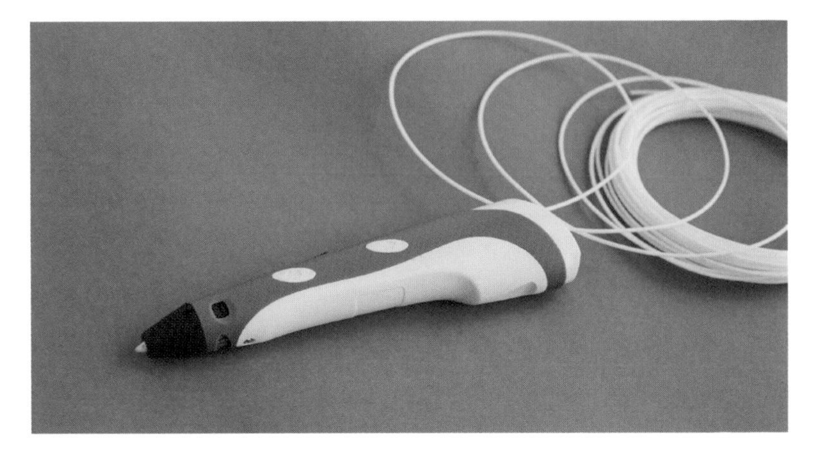

The 3D printing pen I purchased was a very simple device. You plug it in, turn it on, insert the plastic filament, and then press a button to melt the filament and extrude it out the nozzle end. Testing it out, I found it worked as advertised.

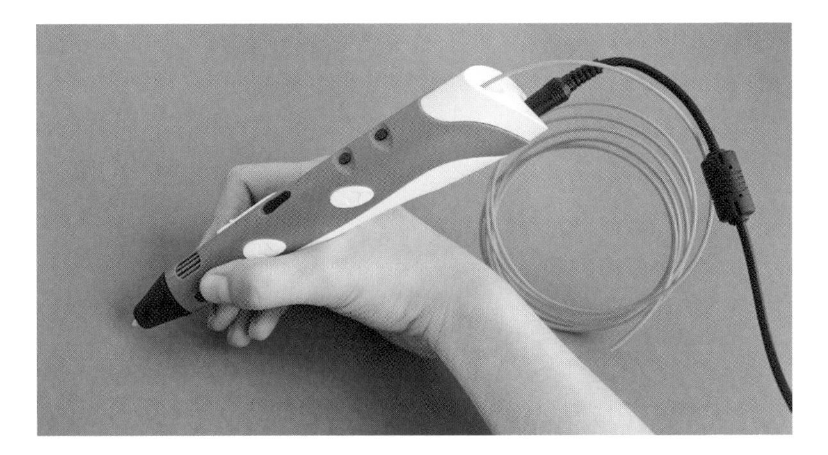

Now I just need to find a way to control it with the Arduino. Since it requires only the press of a button to melt and extrude the plastic, if we can trigger that button with the Arduino, then we should be able to program when it extrudes the plastic and when it doesn't.

What we want to do is switch the button on and off using an electrical signal. The best tool for doing that is a transistor, a device which switches electrical power off and on. In this case, I'll be using an NPN transistor. Taking apart the 3D pen, I found where the extrude button was connected to the main circuit board. We can connect the Arduino to the NPN transistor, and then connect the NPN transistor to either side of the button. Since the Arduino outputs five volts of power and that might be too much power for the button, I added a twenty-two ohm resistor to limit the power. Depending on what 3D pen you use, the resistance may change.

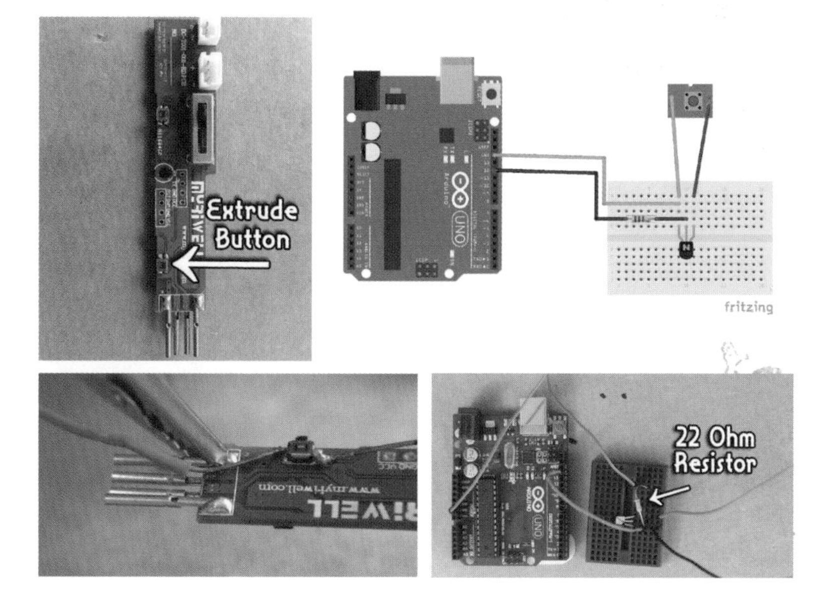

After reassembling the 3D pen and making a space for the wires to run through the casing, I mounted it to the Z axis sliding motor tray. You could use wire, bolts, and spacers, but what I ended up using was hot glue to hold it into place. Now, we've got the bones of our 3D printer! Next, we need to make it print something.

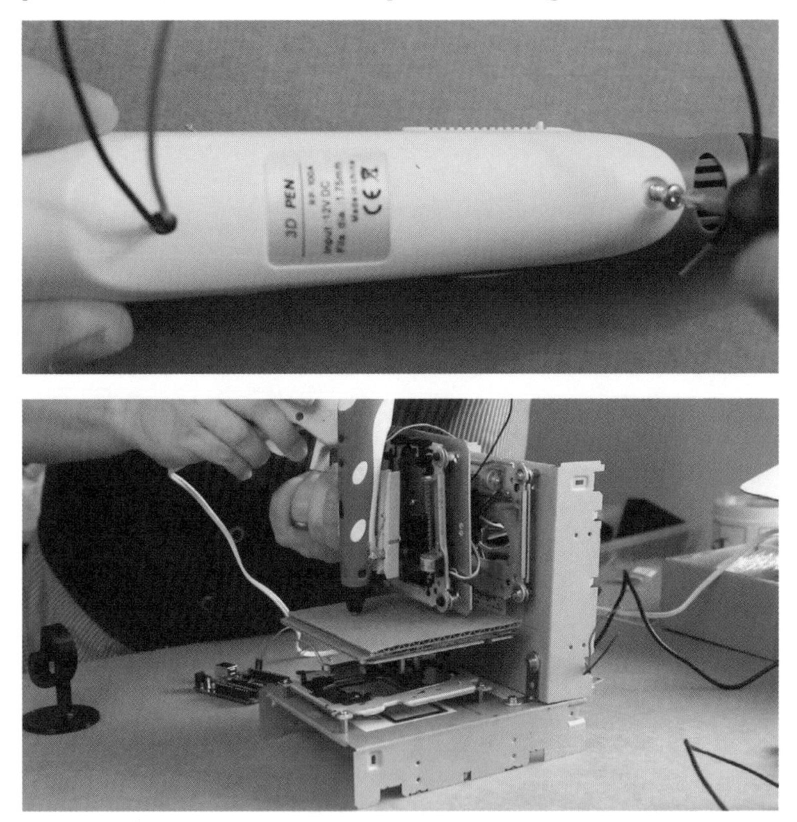

Step 6: The Software

This 3D printer, as well as most other 3D printers and CNC machines, runs off of a programming language called "G-code." It essentially tells the X, Y, and Z axis which specific coordinates it needs to go to in order to make lines and arcs. G-Code is how a 3D printer knows what to create. If you want to 3D print a cube, the instructions for how to build that cube are converted into G-code and sent to the 3D printer. By itself, Arduino has a difficult time interpreting G-code, so we will need to install a G-code interpreter program called "Grbl." You can download the version of Grbl *specific to your Arduino model* from GitHub[4] and load it to your Arduino using a utility called "Xloader"[5].

The Arduino is now ready to accept G-code and send those instructions to our 3D printer motors. The final piece of software we need is something that takes a 3D G-code file and sends those commands to the Arduino. I stumbled across a simple Windows only program called "Grbl Controller."[6] After downloading and installing the program, connect your Arduino to your computer and launch. You can select your Arduino COM port number from the "Port Name" drop-down box, and then click "Open" to connect to it.

Now you can use the arrows on the lower right to jog through the motors. The drop-down box in the lower right corner sets the movement speed. Make sure the speed setting is set to one instead of ten. If any of the platforms are running backward, you can go to Tools > Options and then invert the axis for the backward motor.

4 https://github.com/grbl/grbl
5 http://russemotto.com/xloader/
6 https://www.tinkernut.com/demos/383_cnc/grblcontroller.zip

Once everything is powered up and running, the final step is finding something to print. Since this 3D printer is very specialized, I had to create the G-code by hand, and you can download the samples I made to 3D print on your own machine[7]. If everything goes well, you should see all the motors move and the 3D pen will start creating 3D objects!

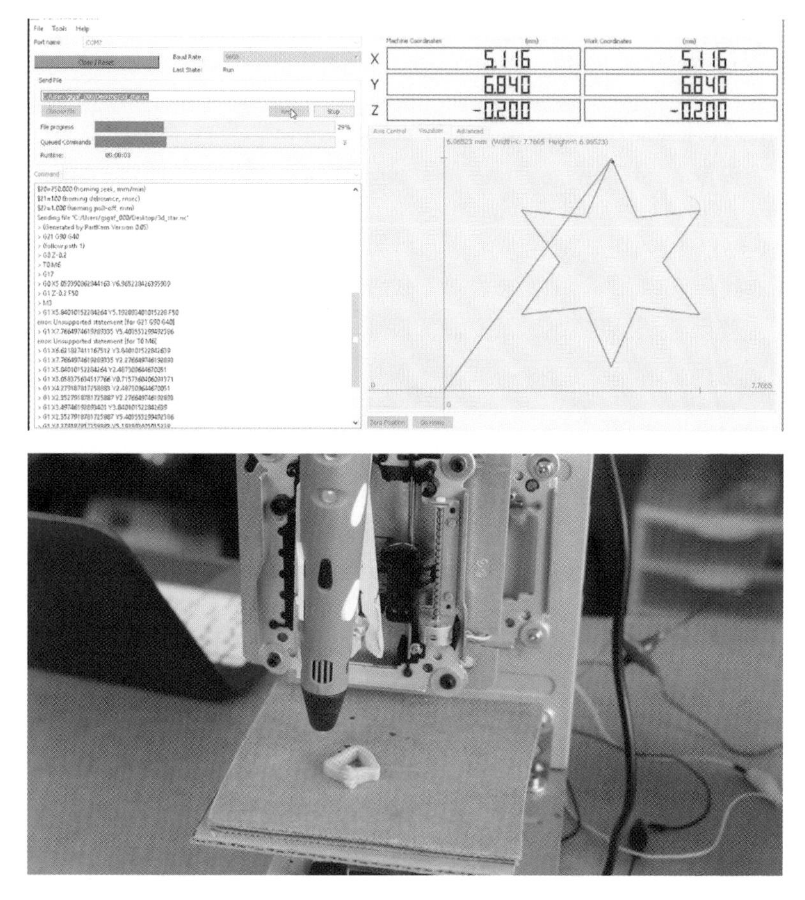

7 http://www.tinkernut.com/demos/384_3d_printer/3d_test_prints.zip

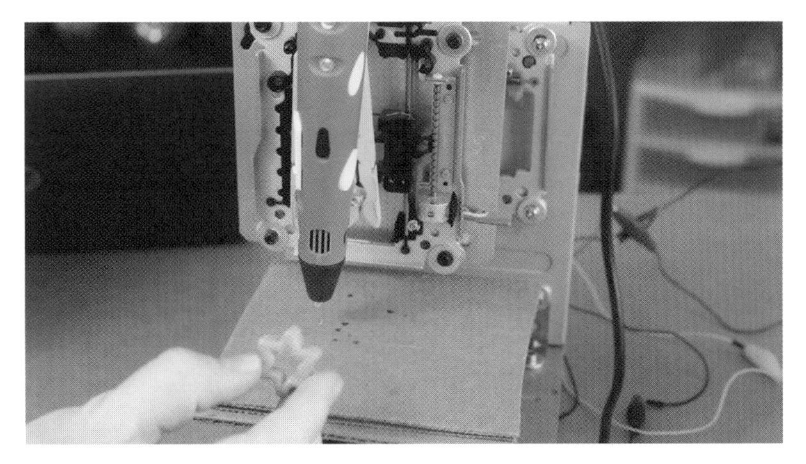

What can I say about smartphones? They're portable and pocket-sized, they've revolutionized the way we communicate and connect with each other online, they've made it easier to document our lives through photo and video, and they entice us to upgrade by coming out with something better and faster every year. Because of that last point, there are so many smartphones lying around going to waste. Although smartphones are currently the predominant mobile device, that wasn't always the case. Believe it or not, older mobile devices didn't connect to the internet. Instead of having everything on one device such as contact lists, music players, and ability to call, these activities all used to be on separate devices. Anyone remember pagers, flip phones, PDAs, and stand-alone MP3 players? Yeah, those were the days.

The great thing about upcycling old mobile devices is that they are... mobile. This means that they are great for making portable projects. The older mobile devices generally have to be modded or taken apart to make something new, but smartphones, especially Android phones, can easily be repurposed just by adding or creating new useful apps! The following chapter has three projects ranging from easy to advanced, and it centers around upcycling different types of mobile devices.

dissentiet vim.

Invidunt neglegentur cu per, labores gubergren voluptaria id duo. An choro maiestatis consequuntur pri. Homero munere nominavi eum no, an sit idque doctus tincidunt. Saepe legimus euripidis cum ut. Dicit tation scripserit vim no, illum commune scriptorem nec in. No vis dicam partem.

eu. I
ri, ea
petua pericula
Usu ut
endo
en hendrerit

n voluptaria
Homero munere
. Saepe
rit vim no,
partem.

MOBILE DEVICES

▶ Difficulty

PROJECT 1

OLD SMARTPHONE TO SECURITY CAMERA

Synopsis: You don't need to pay for a home security system to get good quality surveillance. All you need is a spare smartphone and an internet connection.

Old Smartphone to Security Camera - Larger images can be found on page 120 of photo glossary

Parts & Tools Needed

- Old Android smartphone

Step 1: Updating Your Old Phone

Lots of people upgrade their phones on a two-year basis. This leaves a lot of older smartphones that are either traded in for a discount, gifted to a friend, or sitting around waiting for the chance to be upcycled. For this project, we're going to focus on older Android phones. Why Android and not Apple? The primary reason for choosing an Android phone over Apple is because the Android operating system is "open source," while iOS on Apple is "closed source." This means that anyone can edit, modify, update, and upgrade the Android software and post it online, while Apple is the only entity that can edit or modify iOS. So, if Apple decides to stop upgrading the software for a specific phone, then there's not really any other way to keep that phone updated.

In contrast, there are several different upgrade options for old Android phones to keep them up-to-date. Depending on the phone, you could install Ubuntu Touch, Resurrection Remix OS, MIUI (developed by Xiaomi), or Lineage OS. Lineage OS supports the largest number of devices.

What we want to do first is update our Android phone. The reason we want to update the phone with new software is both because it's more secure and because it increases the compatibility with newer apps in the app store. If your phone is new enough that the manufacturer still pushes out updates to it, then you are good to go.

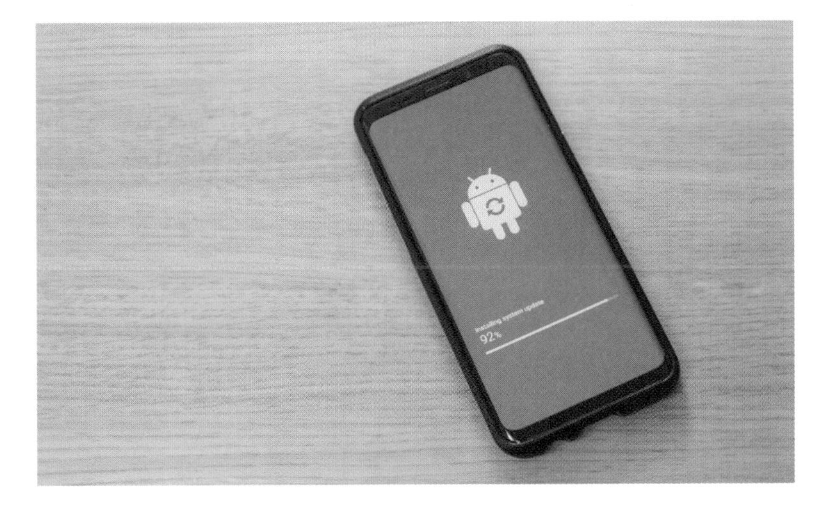

But if your phone has been abandoned by the manufacturer, then you can install a third-party version of Android such as Lineage OS[8]. Lineage OS is the most popular third-party version of Android, and they have a good list of supported devices sorted by manufacturer[9]. Once you have found your phone, clicking on the link will take you to a page that lays out how to install it.

Step 2: Creating a Mobile Security Camera

Once your phone is up to date again, the next step is to install an app that will let us use the phone as a security camera. There are plenty of apps in the app store that can achieve that, but please be

8 https://lineageos.org
9 wiki.lineageos.org/devices

careful when downloading unknown apps to your phone. In rare circumstances, the app store could contain malicious apps that can install viruses on your device. A good app that is trusted and full-featured is called IP Webcam[10].

The free version contains ads, but otherwise it's fully functional. Once installed, select the "Open" button to open it. It opens with a menu of different options and tweaks that you are welcome to skim through, but to get it started right away as a webcam, scroll to the very bottom and select "Start Server." This will launch the camera interface.

10 https://play.google.com/store/apps/details?id=com.pas.webcam

Step 3: Connecting to the Security Camera

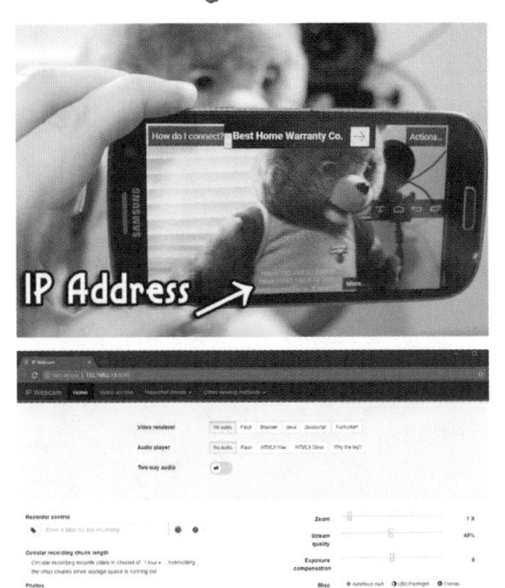

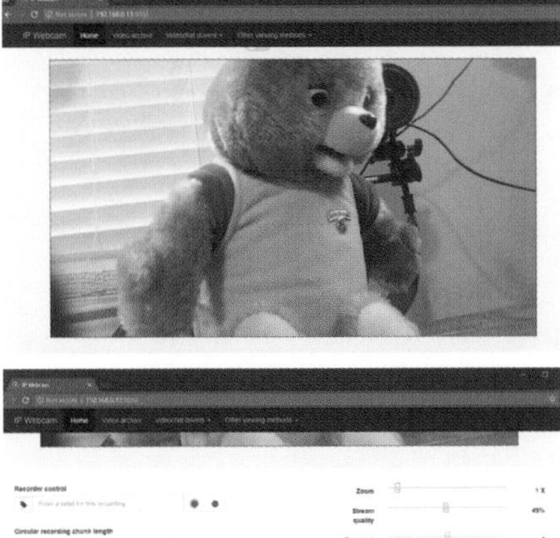

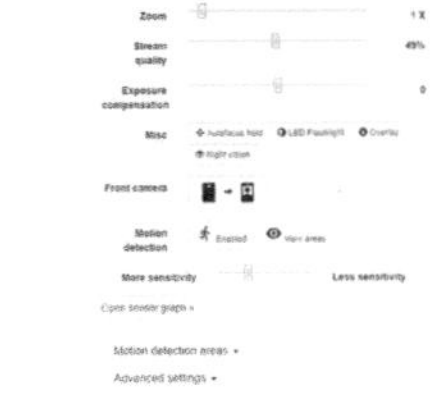

If everything starts up correctly, you'll see the video feed on your mobile screen. At the bottom of the screen you'll see a set of numbers called an "IP address" and a port number. The purpose of this IP address is to allow other devices to connect to the mobile security camera. Now, on a different device that's connected to your network, open up a browser and type in the IP address from your mobile device. It's important for the other device to be on the same network as the mobile security camera. On the web browser, you will see buttons for remotely controlling and viewing the security camera. To view the camera in your browser, you can click on the button that says "Browser" next to the "Video Render" header. Other buttons will let you control features like turning on the LED light, taking recordings, taking photos, and switching from the back camera to the front camera.

With this setup, you can access the security camera from any device connected to your network. If you would like to access the security camera from outside your network, it requires setting up a "Port Forwarding" entry on your network's router. How you can access the router varies depending on which router you have, but in general, you can open up a web browser on any device connected to your network and type in either 192.168.0.1 or 192.168.1.1 into the address bar of your browser. In most cases, this will bring up your router with a login page.

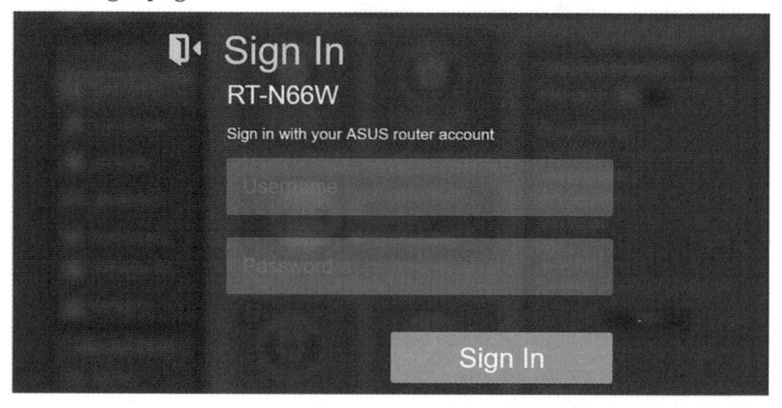

If you haven't changed the login information, it can be found either on the bottom of your router itself or by searching Google for the default login for your router.

Once you have logged in to the router, you can skim through the different settings for your router until you find an option labeled "Port Forwarding" (though some routers may not have this option).

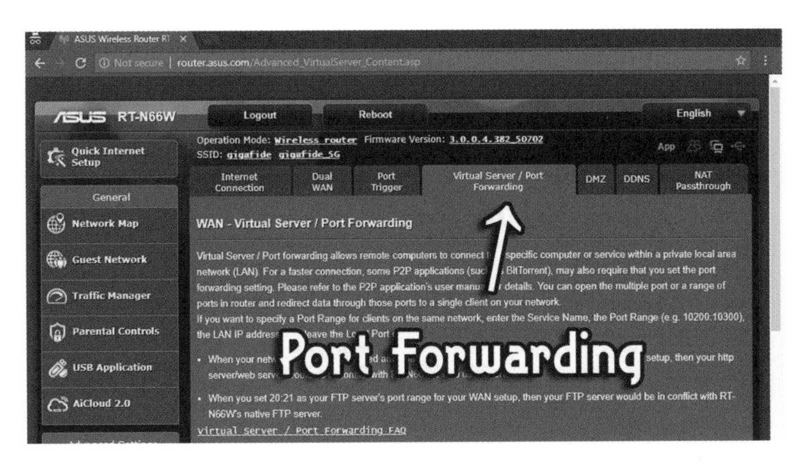

Under this option, you can create a new entry and enter the IP address (for example, 192.168.0.13). It should also ask for a port range and/or a local port, which is where you should put the port number (for example, 8080). Another option defines which "Protocol" to use. Most routers have these options: TCP, UDP, Both. The camera uses the TCP protocol, so selecting that option should work.

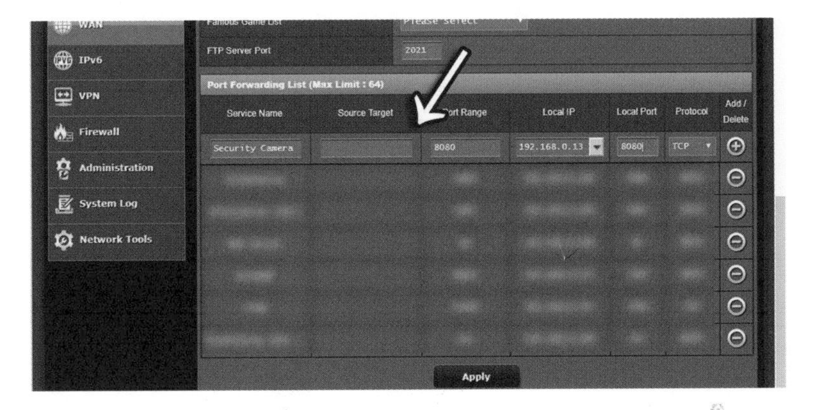

In order to access the webcam externally, the final step is to find the external (or public) IP address for your internet connection. To find the external IP address, open up a browser from any device on your network and then search for "what is my IP address" using the Google search engine. (Note that this may not work using other search engines.)

Google what is my ip address

All News Books Shopping Videos More Settings Tools

About 310,000,000 results (0.28 seconds)

Your public IP address

Learn more about IP addresses

Feedback

What Is My IP Address - Everything You Need to Know
https://whatismyipaddress.com/
Find out what your IP address is revealing about you! **My IP address** information shows your city, region, country, ISP and location on a map. Many proxy servers ...
Ip Lookup · WAN IP Address. LAN IP · Hide My Ip · What is a Private IP Address?

What Is My IP? Shows your real IP - IPv4 - IPv6 - WhatIsMyIP.com®
https://www.whatismyip.com/

The Google results should display your external IP address. With this information, you should be able to get on any network connection anywhere, open up a web browser, and then in the URL bar, type in your external IP address followed by a colon and then the port number for your security camera (for example, 99.88.77.66:8080). Assuming everything worked correctly, this should bring up your security camera's interface!

PROJECT 2

REVIVE AN OLD IPOD

Intermediate

Synopsis: iPods are one of the classic devices that started the Apple renaissance. If you still have one lying around, why not give it a much-needed hardware and software upgrade?

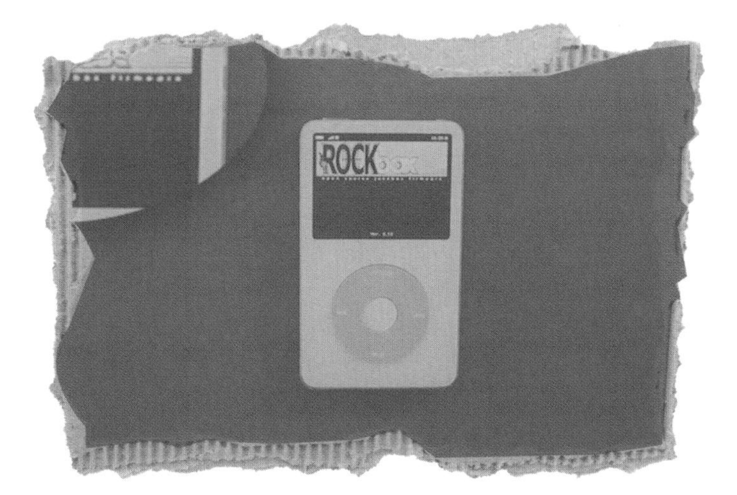

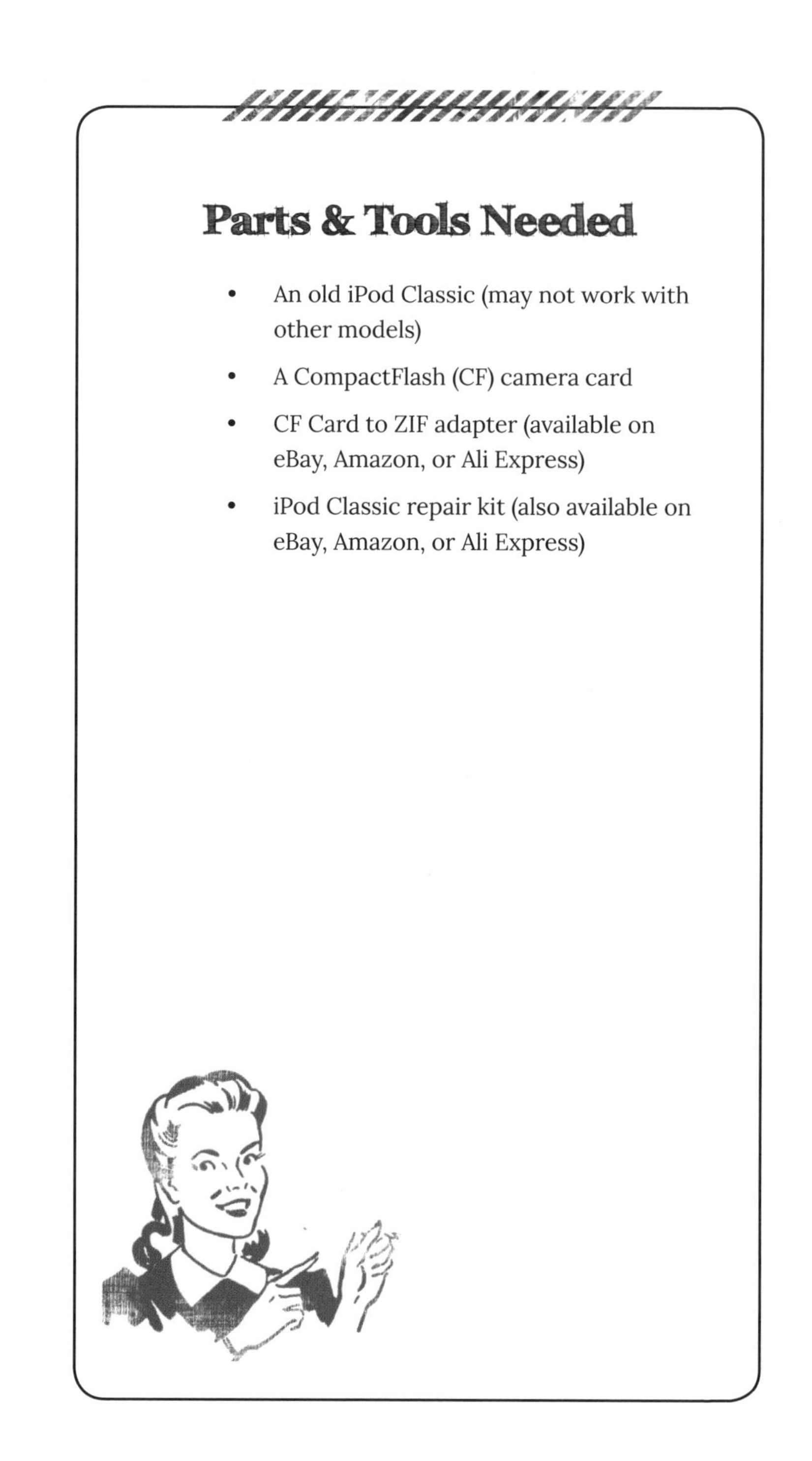

Parts & Tools Needed

- An old iPod Classic (may not work with other models)
- A CompactFlash (CF) camera card
- CF Card to ZIF adapter (available on eBay, Amazon, or Ali Express)
- iPod Classic repair kit (also available on eBay, Amazon, or Ali Express)

Step 1: Taking Apart the iPod

Before internet-connected mobile devices that could stream music to your phone, there were internet-less mobile MP3 players that basically only played music and nothing else. Sounds ancient, right? Out of all the MP3 players in the market, the one that dominated the space was the iPod. The iPod was a huge milestone not only in mobile technology but also in fashion.

It just wasn't cool to own a portable MP3 player that wasn't an iPod, and since I owned several cheap MP3 knockoffs, I was the epitome of "uncool." The only downside to the older iPod Classics came with mechanical hard drives that had limited storage space and were easy to break if the iPod was dropped or jostled around. iPods were notorious for giving users the "frowny face of death."

I was given one of these broken iPods (a fifth generation classic) and wanted to see what I could do to revive it with upgraded, more stable storage.

Before we get started, keep in mind that swapping out the hard drive on your iPod will cause you to lose everything that was on it and will be like starting from scratch. If you are OK with that, then the first step is to crack open the iPod and see what's inside. Maybe I shouldn't use the word "crack," because Apple devices are very fragile and can easily crack or break if you're not too careful. While a skillful hand might be able to pry it open with an eyeglass screwdriver and a guitar pick, I would highly recommend investing in an iPod repair kit to avoid any disasters. Using the repair kit pry tools, find where the front panel meets the metal back casing and slowly begin to pry it open. Be very careful and just go around the perimeter of the iPod prying away the panel a little bit each time. Eventually it will pop off.

Once the front panel is separated from the backing, there are a couple of cables that connect it to the main logic board. The cable that goes to the battery is easy to disconnect. Just find where it connects to the logic board, and there should be a release flap on the connector that disconnects the cable.

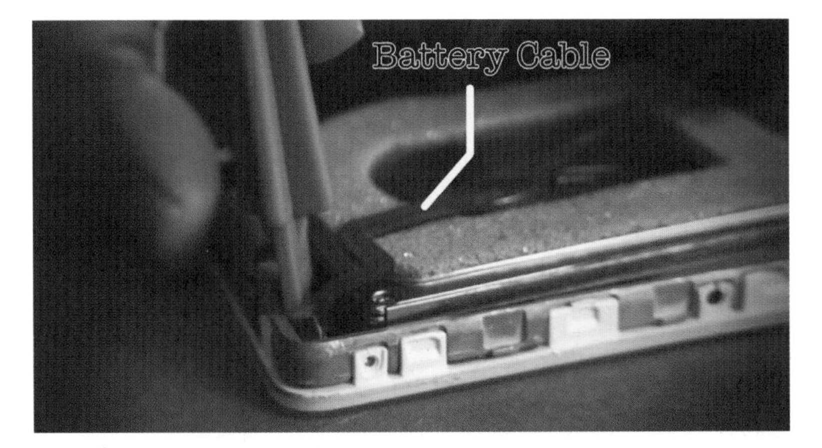

There is one more cable that goes to the iPod hold switch, but I was able to swing open the front panel and lay it beside the back casing without having to remove this cable.

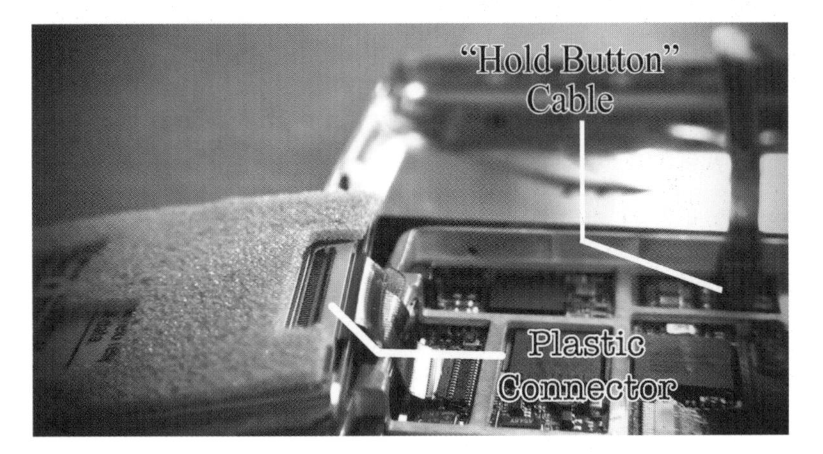

Revive an Old iPod - Larger images can be found on page 122 of photo glossary

Step 2: Replacing the Hard Drive

The hard drive should be surrounded by rubber bumper guards and connected by a small cable called a ZIF cable. You can remove the bumper guards and flip the hard drive downward to expose where the ZIF cable connects to it. Unsnapping the little plastic connector clip will release the cable and allow you to remove the hard drive.

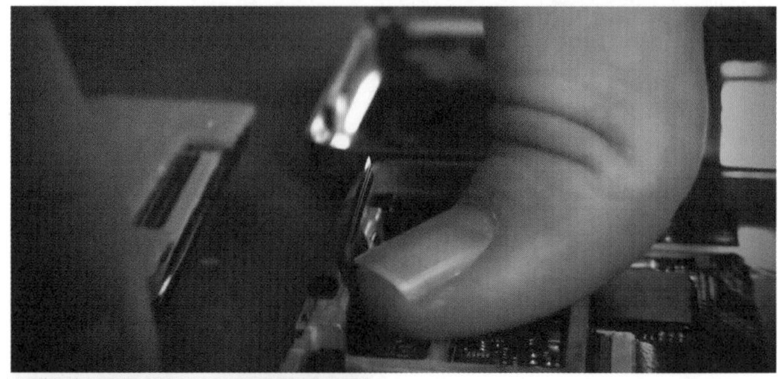

With the hard drive removed, we can now replace it with the CF card. CF cards come in a range of storage sizes up to 256 gigabytes at the time of this book's writing. The first step is to place your CF card in the adapter. There should be a plastic connector on the adapter that the ZIF cable attaches to. Connect the cable to the adapter, make sure it's secure, and then position it inside the iPod where the hard drive used to be. Now that everything is connected, reconnect the battery cable and replace the front panel.

Step 3: Reinstalling the Software

When you first turn the iPod on, it doesn't know how to recognize the blank CF card, so the screen will tell you to connect the iPod to your computer to restore it.

Restoring an iPod requires iTunes and an Apple account. If you don't already have iTunes, it's free to download and install on your computer. Assuming iTunes is installed on your computer, plug your iPod into it. It should be autodetected by iTunes, at which point it should give you an option to restore it. Clicking the restore button will reinstall the iPod software onto it. Then you can load up your music and start using it as an iPod again!

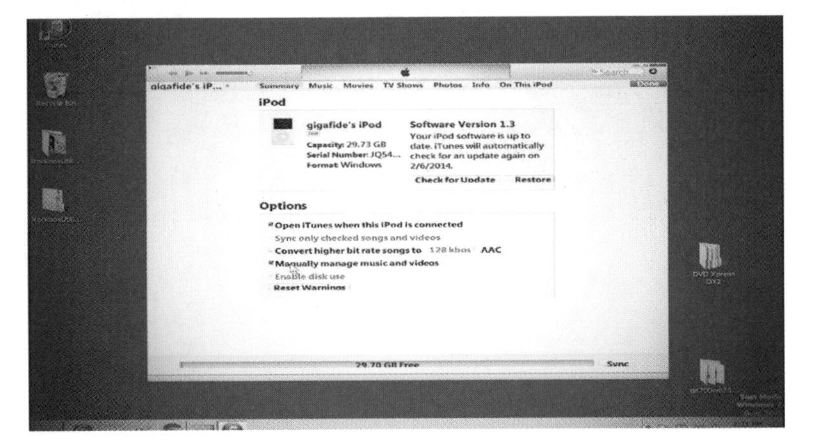

If you want a more customized experience that the default Apple software can't give you, you can also install Rockbox (www.rockbox.org). It's an alternative operating system for iPods and other MP3 players. Rockbox free, and it allows you to upload custom fonts, themes, icons, and even play Doom! It's simple to install. Just download the Rockbox utility, plug your iPod in, launch the utility, and click install!

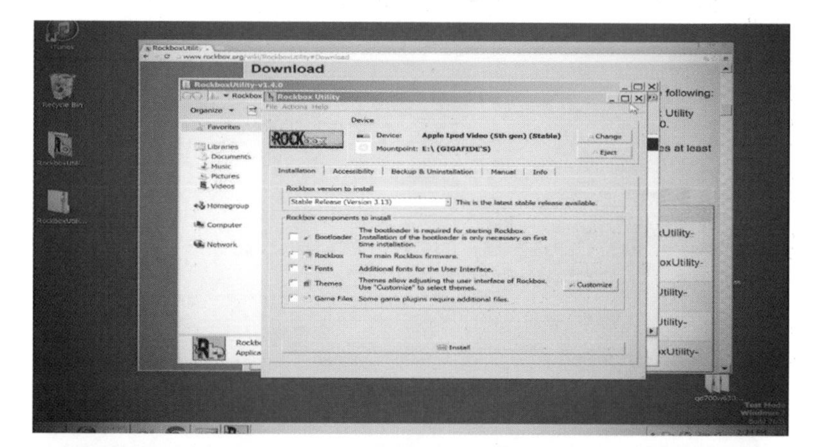

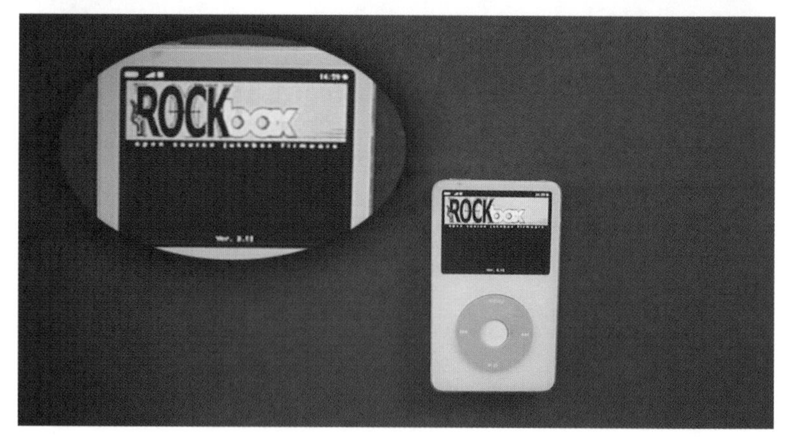

PROJECT 3

OLD CELL PHONE TO SMARTWATCH

Advanced

Synopsis: Before there were smartphones, there were "feature" phones. Remember? They were the ones with physical buttons and monochrome screens. Still got one of those lying around somewhere? Here's how you can turn that old phone into a smartwatch!

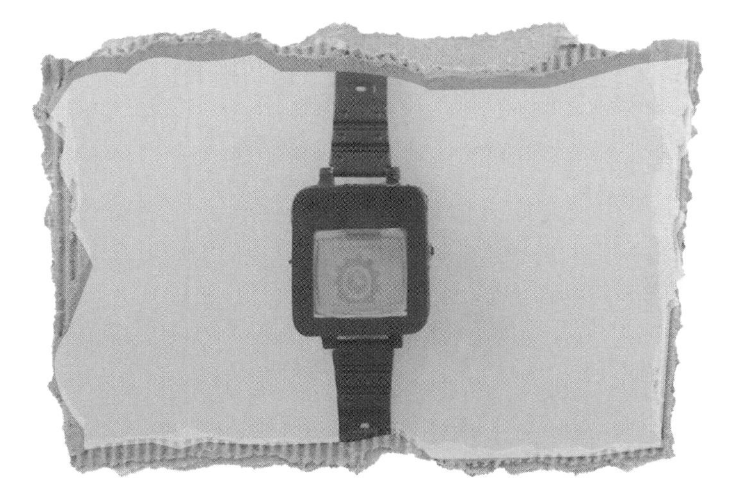

Old Cell Phone to Smartwatch - Larger images can be found on page 124 of photo glossary

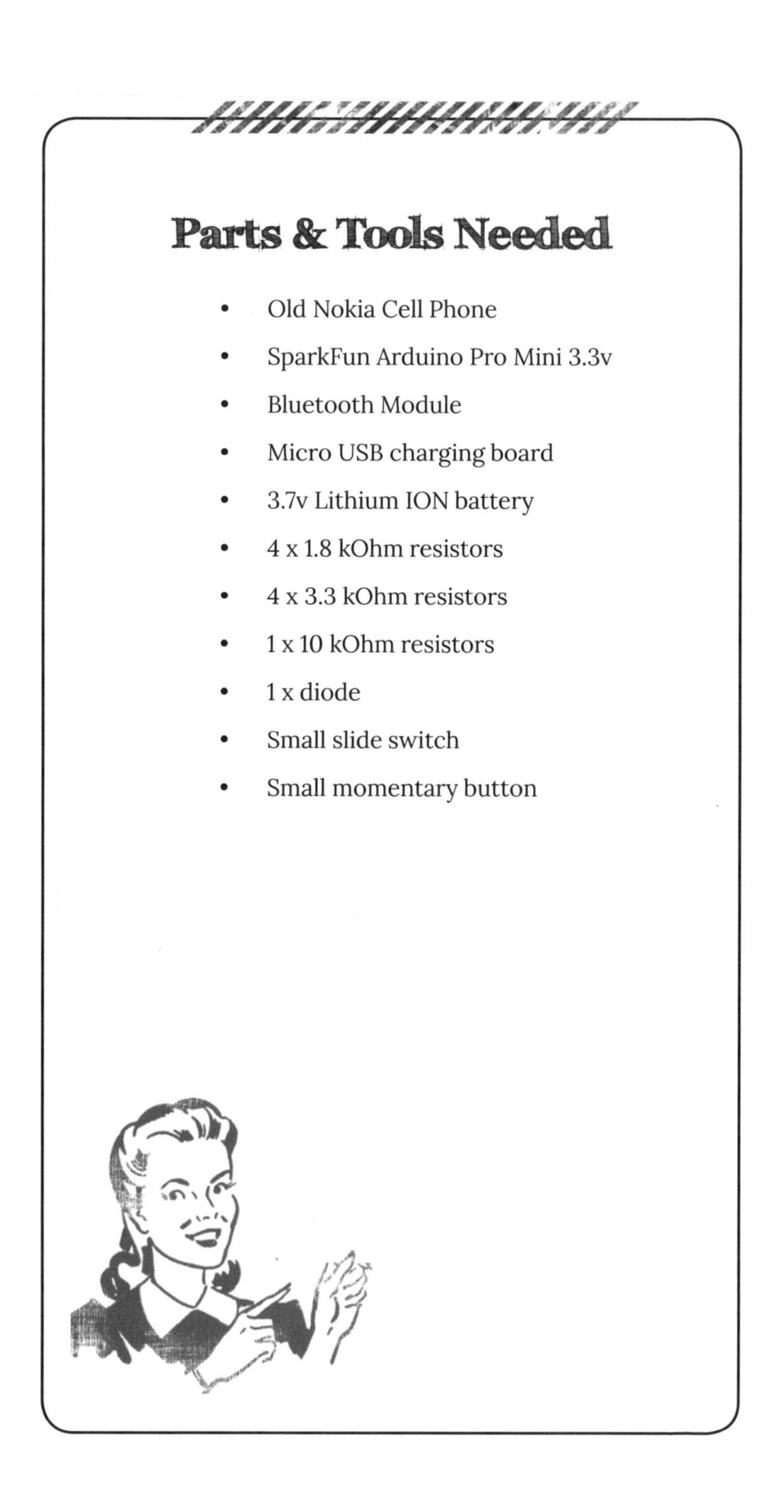

Parts & Tools Needed

- Old Nokia Cell Phone
- SparkFun Arduino Pro Mini 3.3v
- Bluetooth Module
- Micro USB charging board
- 3.7v Lithium ION battery
- 4 x 1.8 kOhm resistors
- 4 x 3.3 kOhm resistors
- 1 x 10 kOhm resistors
- 1 x diode
- Small slide switch
- Small momentary button

Step 1: Scavenging Parts

What I'd like to do for this crazy/ambitious project is turn an old cell phone into a smartwatch. The primary reason for this project is simply that I had an old cell phone lying around and wanted to find a creative way to repurpose it. By "old cell phone," I mean the ones that had the small LCD screens and physical keyboard/buttons. The one I had is a Nokia 1100, but almost any other old cell phone would work, so long as you can find the schematics and information for it online.

The first thing we want to do is take the phone apart and see what we can scavenge from it. The most important item we want to try and scavenge is the screen.

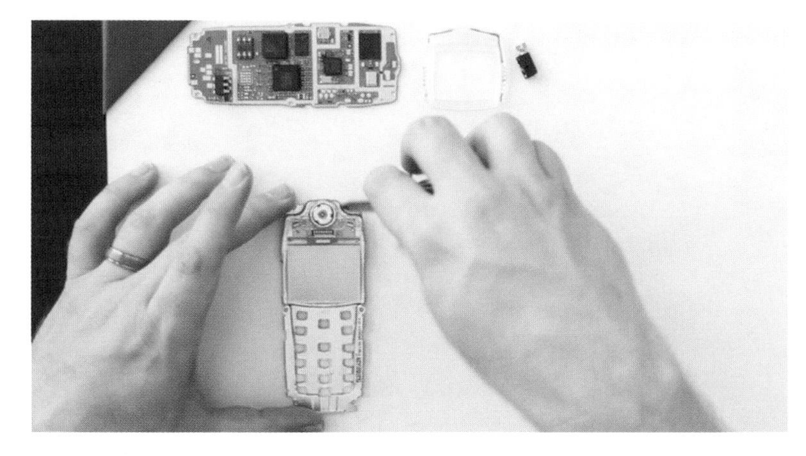

From my Nokia, I was also able to get a vibrating motor and a speaker, which wasn't as much as I was hoping, but was at least a good start.

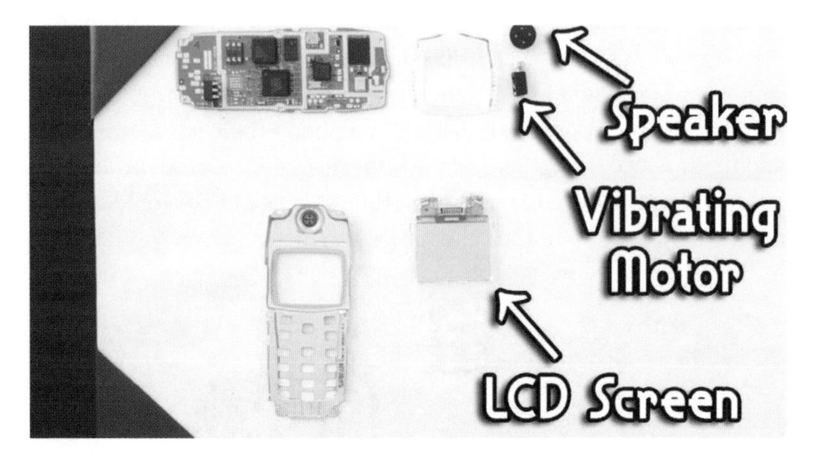

If you're not sure how to extract the parts or if your LCD is salvageable, do a quick Google search to see if you can find some teardown information for your phone and schematics or a pin-out diagram for the LCD screen. What you're able to scavenge really depends on what type of phone you have and how old it is. The older the phone, the more scavengeable parts you will find.

Given what I was able to extract, the parts I can use in the watch are the screen and the vibrating motor. While it would be nice to use the speaker to add audio feedback to the watch, it's a little too bulky for this project. Now that we have a couple of core parts, we can start building a watch around them. Let's begin with the LCD.

Step 2: Figuring Out the LCD

Without a working LCD, we have no project, so we need to figure out how to get our LCD working outside of the phone. The first step is to see if I can find a schematic for it online. Doing a quick Google search for "Nokia 1100 LCD," I was able to find out a ton of information about it. I found that the LCD model number is PCF8814. I was also able to find a schematic that listed the pinouts (what each of the pins on the LCD represents).

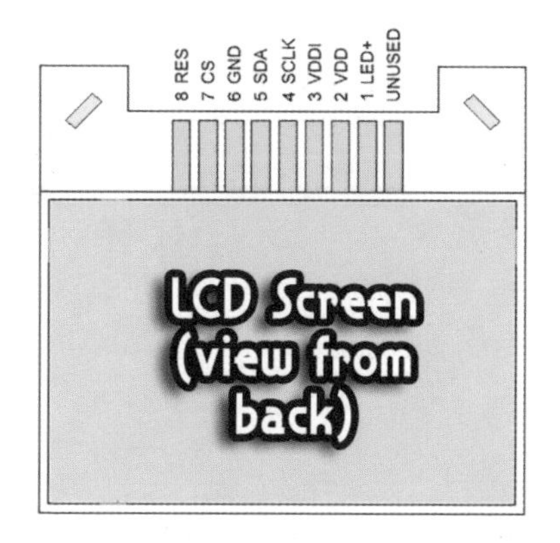

This is a view from the back side of the LCD, so pin eight is on the far left when looking at the back. There are nine pins, but one is unused, leaving eight usable pins. The pins on the LCD are very small, so in an effort to make them easier to connect to, I soldered wire from a ribbon cable to each pin, and then hot-glued it all into place. So now I can connect each of the wires to a breadboard to make it easier to work with the LCD.

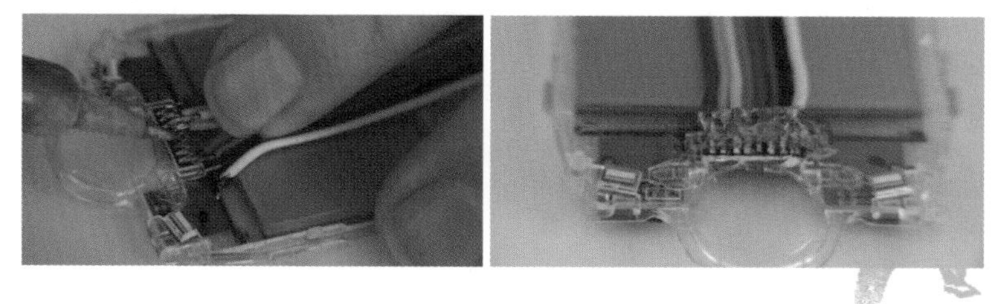

Step 3: Driving the LCD

Now that I have an LCD screen ready for reuse, I need something that can control or "drive" it. Generally, a microcontroller is a good choice to drive an LCD, and there are lots of microcontroller options available. I decided to go with an Arduino Pro Mini because it's very small and there is a fantastic Arduino community online for support. Another benefit of using an Arduino is that I can use a larger Arduino Uno for prototyping and then switch to the Arduino Pro Mini when I'm ready to shove it into a smartwatch.

Let's start wiring stuff up! First, I connected all the LCD wires to a breadboard, and then I began connecting wires to the Arduino Uno. (I'll switch all connections to the Arduino Pro Mini later when I'm done testing.) I added some resistors to the pins that connect directly to Arduino pins to keep the Arduino from sending too much power to the LCD and burning it out. You can use the diagram for reference, but it ended up being a total of four 1.8k Ohm resistors and four 3.3k Ohm resistors.

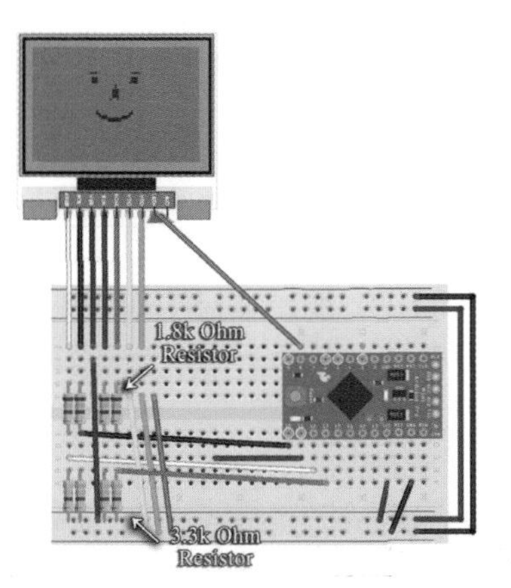

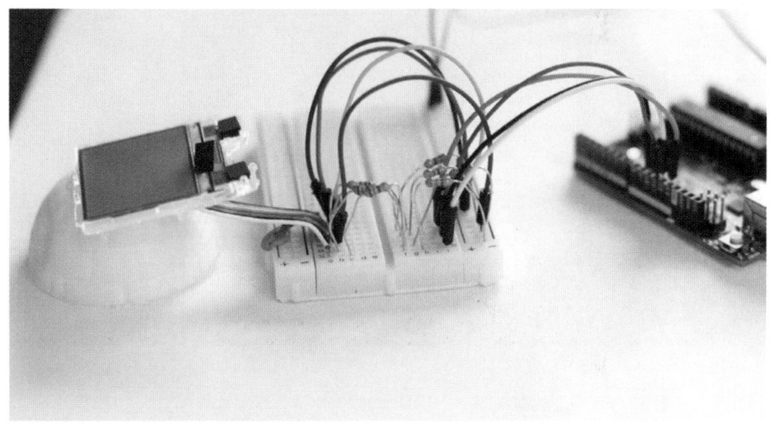

If you were to power up the Arduino at this point, you should see the backlight come on for the screen, but nothing will display on it until we add some code, so let's do that next.

We have the LCD connected to our Arduino driver, so the final key to making the LCD useful is to program some code into the Arduino that sends text to the LCD. Arduino has its own programming software that you can download from their website.[11] After installing software, plug your Arduino into your computer and launch it. If I were to write the entire code from scratch, I would have to tell each Arduino pin what to display, how to display it, when to turn off, etc., which would take forever. Luckily this can be resolved through the use of "libraries"[12]. I was able to find a couple of different libraries for the PCF8114, but I chose this one from GitHub user cattzalin[13] because of its ability to display bitmap images.

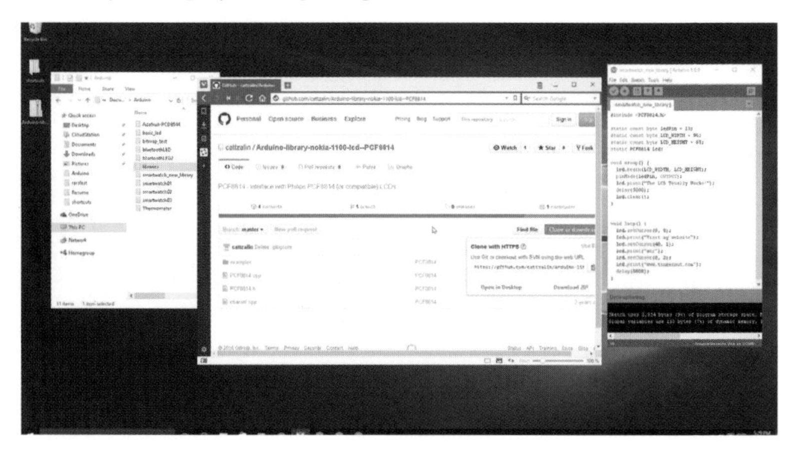

You can download it, unzip it, and then move it to your Arduino libraries folder (check out this guide[14] for more information). You can then open up the Arduino software and start writing some code. You can use my code below as reference, but basically I imported the library, set the variables for the type of screen I was using, and then I sent some text to the screen.

11 https://www.arduino.cc/en/Main/Software
12 https://www.arduino.cc/en/Reference/Libraries
13 https://github.com/cattzalin/Arduino-library-nokia-1100-lcd--PCF8814
14 https://www.arduino.cc/en/Guide/Libraries

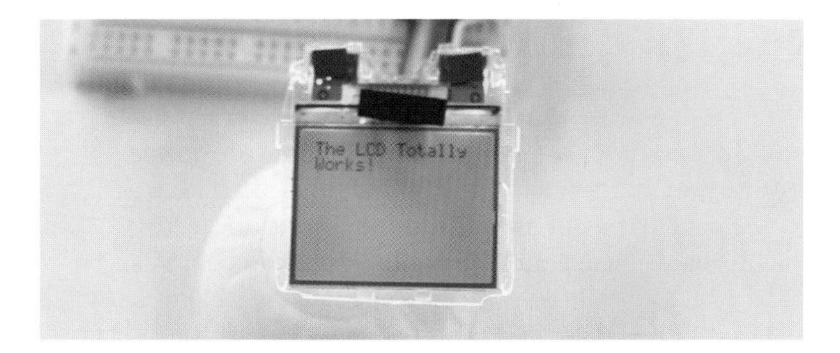

Plug in your Arduino, upload the code, and if everything works, you should see your text on the screen!

Now that we have our LCD functioning, we can move on to the next step!

```
#include <PCF8814.h>
static const byte ledPin = 13;
static const byte LCD_WIDTH = 96;
static const byte LCD_HEIGHT = 65;
static PCF8814 lcd;
void setup() {
        lcd.begin(LCD_WIDTH, LCD_HEIGHT);
        pinMode(ledPin, OUTPUT);
        lcd.print("The LCD Totally Works!");
        delay(5000);
        lcd.clear();
}
void loop() {
        lcd.setCursor(0, 0);
        lcd.print("Visit my website");
        lcd.setCursor(40, 1);
        lcd.print("at:");
        lcd.setCursor(0, 2);
        lcd.print("www.tinkernut.com");
        delay(5000);
        }
```

Step 5: Adding Bluetooth

Right now, all we have is an LCD screen on which we can display stuff. As a watch, it's nothing special. Every watch can do that. The magic dust that will turn this LCD screen into a smartwatch is Bluetooth. Bluetooth is a wireless standard that will let the smartphone communicate with the Arduino and to send it data like time, date, notifications, etc. I ended up going with a JY-MCU (HC-06) Bluetooth module. It's about the size of the Arduino Pro Mini. There are smaller options available, but they require surface mount soldering skills, which I don't really want to get into for this project.

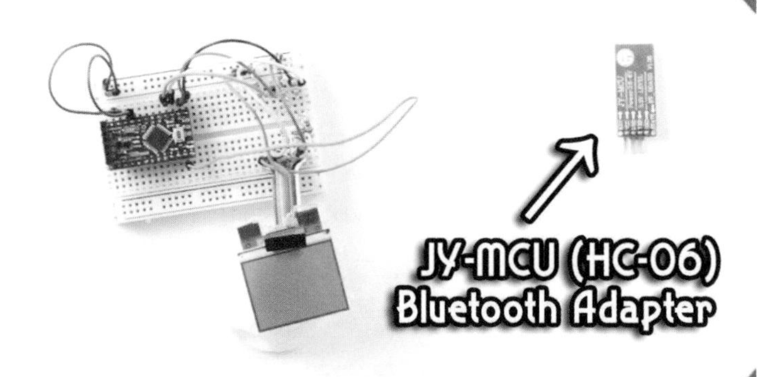

Connecting it is pretty simple. Here's a diagram:

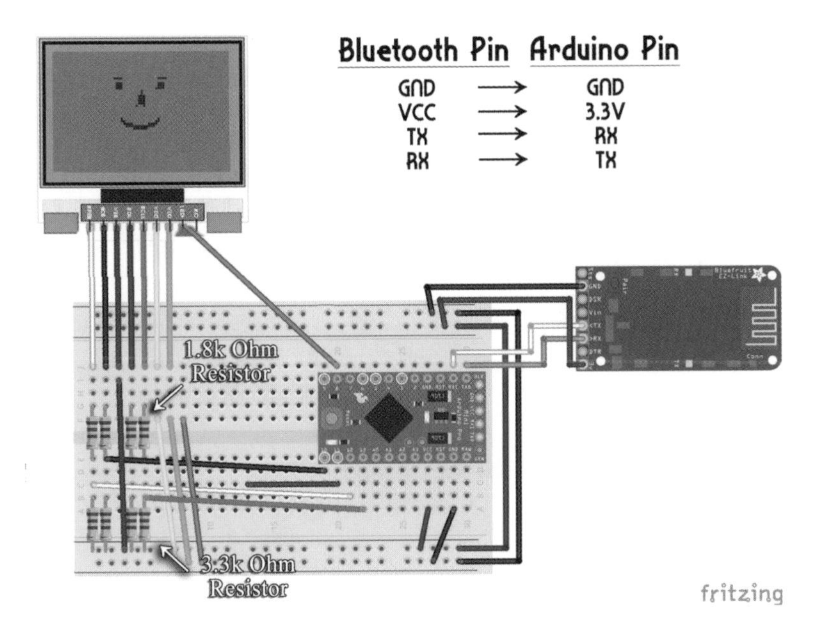

Step 6: Other Components

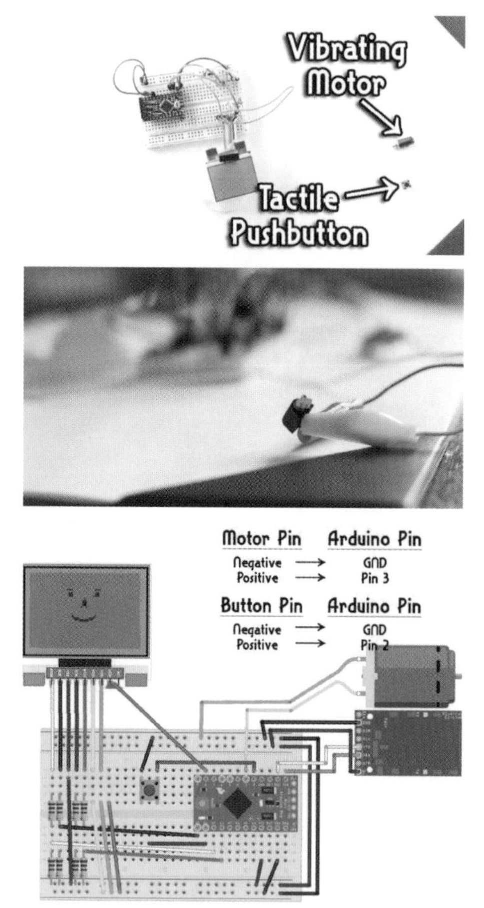

There were a couple more hardware features that I wanted to add to the smartwatch: a vibration motor for notifications and a button to turn the backlight on and off. To properly wire a DC motor to an Arduino, it normally requires a transistor, a diode, and a resistor to avoid burning out the Arduino and/or the motor. But since this is a very small motor with very small voltage, and because there is very little space for it, I'm just wiring it directly to the Arduino. The last element I added was a momentary switch button wired to pin two and the GND pin of the Arduino.

Step 7: Updating the Arduino Code

The Arduino code that we've written so far simply displays characters on the LCD, but we want it to do a lot more than that. The first thing I want to add is a "splash" screen, which is basically a logo or image that is displayed when the watch is booting up. You can't insert an image into Arduino code like you would insert an image into a Facebook post. Instead, we have to convert the image into code that can be read by the Arduino. To create the coded version of an image, you'll need a black and white image, and you'll need to make sure it's sized to within the resolution of your screen and saved as a bitmap file format. For the Nokia screen I'm using, it's ninety-six by sixty-five pixels. Then you can download a free image conversion program called LCD Assistant[15]. Just load the image into the program and save the output as a text file.

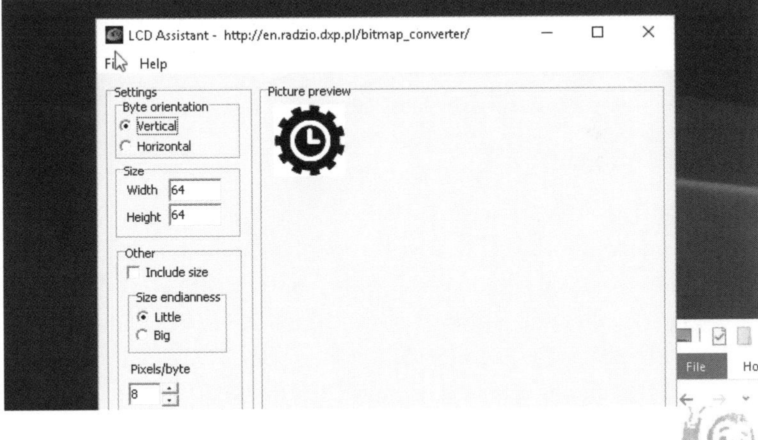

When you open up the text file, you will see the image code that you can use in your program.

15 http://en.radzio.dxp.pl/bitmap_converter/

```
//------------------------------------------------------------------------
// File generated by LCD Assistant
// http://en.radzio.dxp.pl/bitmap_converter/
//------------------------------------------------------------------------

const unsigned char tinkerwatch_logo [] = {
0x00, 0x00, 0x00, 0x00, 0x00, 0x00, 0x00, 0x00, 0x00, 0x00, 0x00, 0x0
0x20, 0xE0, 0xF0, 0xF0, 0xF8, 0xF8, 0xF8, 0xFC, 0xE0, 0xC0, 0xC0, 0xC
0xE0, 0xFE, 0xFE, 0xFE, 0xFE, 0xFE, 0xFE, 0xFC, 0x8C, 0x80, 0x80, 0x0
0x80, 0xC0, 0x80, 0x80, 0x00, 0x00, 0x00, 0x00, 0x00, 0x00, 0x00, 0x0
0x00, 0x00, 0x00, 0x00, 0x00, 0x00, 0x80, 0xC0, 0xF0, 0xF0, 0xE0, 0xC
0xF8, 0xFC, 0xFF, 0xFF, 0xFF, 0xFF, 0xFF, 0xFF, 0x7F, 0x7F, 0x3F, 0x3
0x3F, 0x3F, 0x3F, 0x3F, 0x3F, 0x3F, 0x3F, 0x7F, 0x7F, 0x7F, 0xFF, 0xF
0xFF, 0xFF, 0xFF, 0xFF, 0xFF, 0x1E, 0x1C, 0x08, 0x00, 0x00, 0x00, 0x0
0x00, 0x00, 0x00, 0x00, 0x00, 0x06, 0x0F, 0x0F, 0xEF, 0xFF, 0xFF, 0xF
0x7F, 0x1F, 0x0F, 0x07, 0x03, 0x01, 0x00, 0xC0, 0xC0, 0xE0, 0xF0, 0xF
0x08, 0x18, 0x38, 0xF8, 0xF8, 0xF0, 0xF0, 0xE0, 0xE0, 0xC0, 0x80, 0x01
0x3F, 0xFF, 0xFF, 0xFF, 0xFF, 0xFF, 0xFE, 0xFC, 0xF0, 0xE0, 0xF0, 0xF
0x00, 0x00, 0xF0, 0xF0, 0xF0, 0xF0, 0xF0, 0xFF, 0xFF, 0xFF, 0xFF, 0xF
0x00, 0x00, 0x00, 0x00, 0xF8, 0xFE, 0xFF, 0xFF, 0xFF, 0xFF, 0xFF, 0xF
0x00, 0x00, 0x00, 0x3F, 0x3F, 0x3F, 0x3F, 0x3F, 0x3F, 0x7F, 0xFF, 0xF
0x00, 0x00, 0x03, 0xFF, 0xFF, 0xFF, 0xFF, 0xFF, 0xFF, 0xFF, 0xFF, 0x0
0x00, 0x00, 0x0F, 0x0F, 0x0F, 0x07, 0x0F, 0xFF, 0xFF, 0xFF, 0xFF, 0xF
0x00, 0x00, 0x00, 0x02, 0x1F, 0x7F, 0xFF, 0xFF, 0xFF, 0xFF, 0xFF, 0xF
0xF8, 0xF8, 0xF8, 0xF8, 0xF8, 0xF8, 0xF8, 0xF8, 0xF8, 0xF8, 0xFF, 0xF
0x00, 0x00, 0xC0, 0xFF, 0xFF, 0xFF, 0xFF, 0xFF, 0xFF, 0xFF, 0xFF, 0xF
0x00, 0x00, 0x00, 0x00, 0x00, 0x70, 0xF0, 0xF8, 0xFF, 0xFF, 0xFF, 0xF
0xFE, 0xF8, 0xF0, 0xE0, 0xC0, 0x80, 0x01, 0x03, 0x07, 0x07, 0x0F, 0x0
0x1F, 0x1F, 0x1F, 0x1F, 0x1F, 0x1F, 0x0F, 0x0F, 0x07, 0x03, 0x01, 0x8
0xFC, 0xFE, 0xFF, 0xFF, 0xFF, 0xFF, 0x7F, 0x1F, 0x07, 0x07, 0x07, 0x0
0x00, 0x00, 0x00, 0x00, 0x00, 0x00, 0x00, 0x03, 0x07, 0x0F, 0x07, 0x0
0x1F, 0x1F, 0xFF, 0xFF, 0xFF, 0xFF, 0xFF, 0xFF, 0xFE, 0xFE, 0xFC, 0xF
0xF8, 0xF8, 0xF8, 0xF8, 0xF8, 0xFC, 0xFC, 0xFC, 0xFE, 0xFE, 0xFF, 0xF
0xFF, 0xFF, 0xFF, 0xFF, 0xFF, 0x7C, 0x38, 0x10, 0x00, 0x00, 0x00, 0x0
0x00, 0x00, 0x00, 0x00, 0x00, 0x00, 0x00, 0x00, 0x00, 0x00, 0x00, 0x0
0x02, 0x07, 0x07, 0x0F, 0x0F, 0x1F, 0x1F, 0x1F, 0x07, 0x01, 0x01, 0x0
0x03, 0x7F, 0x7F, 0x7F, 0x7F, 0x3F, 0x3F, 0x3F, 0x39, 0x01, 0x00, 0x0
0x00, 0x03, 0x01, 0x00, 0x00, 0x00, 0x00, 0x00, 0x00, 0x00, 0x00, 0x0
```

Let me add as a disclaimer that my code is very buggy, but I've made it open source. The code is thoroughly commented and documented, so you can copy the code from my GitHub page[16], but here's a brief explanation of what it does. Aside from adding an image to my Arduino code, I also added some code that reads data from the Bluetooth Serial connection. So if data is sent from a connected Bluetooth device (such as a smartphone), it will read that data and process it. The data that it's going to be processing

16 https://gist.github.com/gigafide/e696555da02b013994b3df1a37d340c2#-file-tinkerwatch-ino

from the smartphone will be the current date, time, a phone number value, and a text value. Each of those values is stored as a variable and displayed to the screen. If there isn't a value sent by the smartphone, then it just displays a default text or no text at all. At the end of the code, I have the section to make the button work, as well as a backlight loop to keep the backlight on for a set period of time.

Step 8: Making an Android App

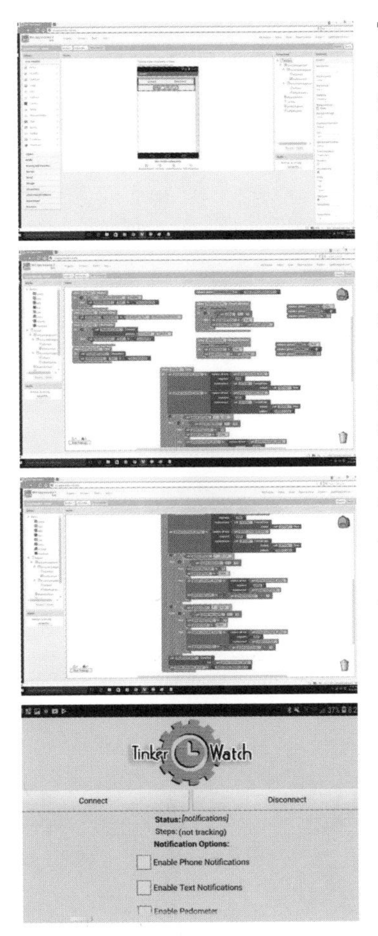

The second half of making this smartwatch "smart" is connecting it to an Android phone. To send data from the Android phone to the smartwatch, we will need to make an Android app that runs on the phone. I know that sounds like a daunting task, but there's a nice cheat for making quick and dirty Android apps. MIT maintains a program called App Inventor[17] that is a simple, user-friendly way of making Android apps. If you want a quick run-through on how to create apps with it, you can watch my "Make An Android App In 7 Minutes" tutorial[18]. The app I made was very simple; it essentially had a connect/disconnect button for the Bluetooth, and then it would send the date/time and phone/text notifications via Bluetooth. Since I can't really export working files from AppInventor, we'll have to settle for images of the layout I created. If you

17 https://appinventor.mit.edu/
18 https://www.tinkernut.com/portfolio/make-android-app-7-minutes/

don't want to create your own code, you can download mine at the Google Play app store. **It's important to note that the Bluetooth device must be paired with the Arduino in the settings before you can use it in the app.**

Step 9: Testing the Functionality

We've got our Bluetooth Arduino watch (connected to a breadboard), and we have our Android app, so let's test it out. After you see the splash screen, it will then wait for input from your phone before displaying time.

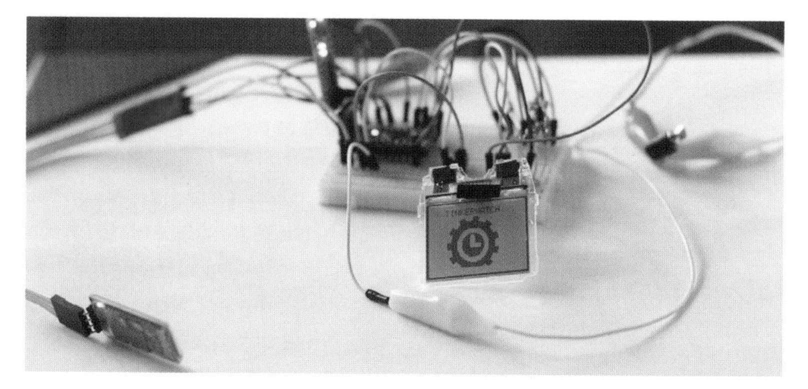

As stated in bold above, make sure that the Bluetooth device is paired with your Android device in the settings first. Then you can launch your app and connect to it. If everything is successful, you should see the date and time on the screen.

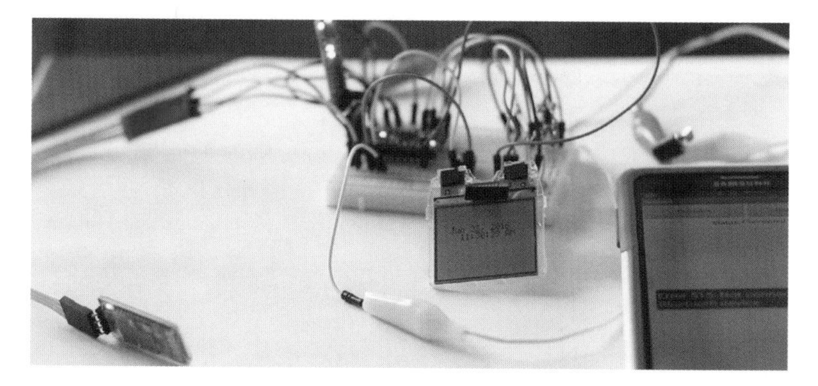

Pushing the button should activate the backlight. And last but not least, whenever you receive a call or a text, the screen should light up and the motor should buzz.

Step 10: Shrinking It Down

Now that we have the software and hardware to make a functioning smartwatch, we now need to make it look like a smartwatch. The first step is to swap out the larger Arduino Uno for the smaller Arduino Pro Mini. Just load the same software onto the Arduino Pro Mini, and then connect all the wires to it in the same way. The next step is to take the mess of wires connected to the breadboard and solder them all together into a simple little circuit board. I ended up adding a diode to the motor for protection and a 10k Ohm resistor to the momentary button. Another tip for the motor is to make sure that the head of the motor isn't being obstructed by anything. Otherwise, the vibrating top of the motor might get stuck.

How you arrange everything is completely up to you, your skill level, and what you have to work with.

There's one key component we're missing. The watch obviously needs a source of power. Ideally, it should be a rechargeable battery that can be recharged using any micro USB cable (just like most portable devices). To achieve this, I got a 3.7v 100 mAh Li-ion battery and a one amp micro USB charging board that acts as a surge protector and regulator.

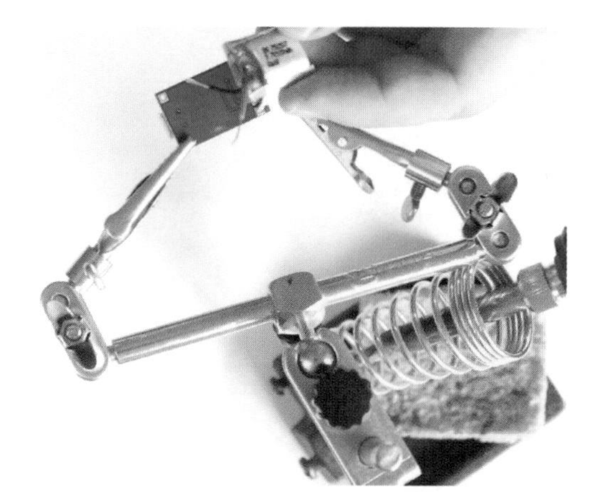

After the project, I found this battery that would probably make a better option. Before soldering the charging board to the Arduino, I tweaked the code and made sure everything was perfect and then reuploaded it one last time to the Arduino. I say "last time" because it will be difficult to upload anything to it after it's all been shoved inside of the smartwatch case. With the code uploaded, you can now

solder the charging board wires to the VCC and GND pins or the RAW and GND pins on the Arduino and then test it out!

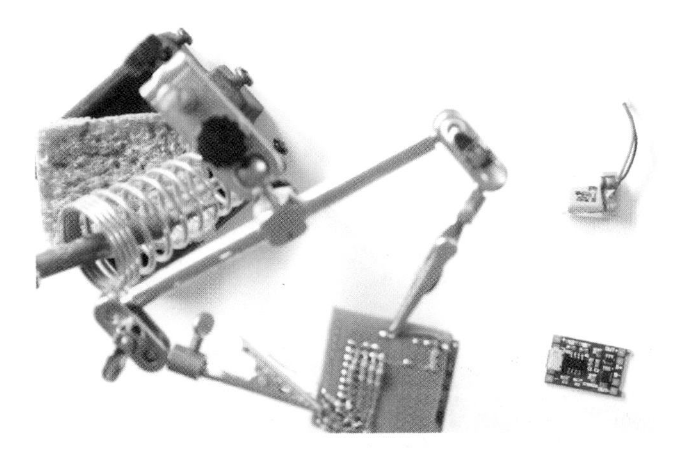

Step 12: Final Steps

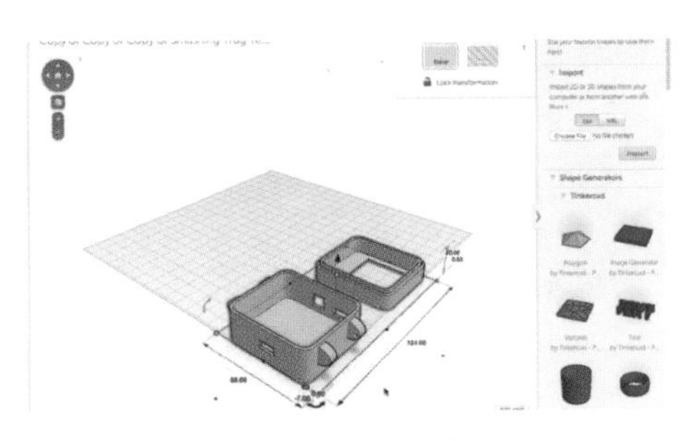

We have made it to the final step! This is the fun part where we take our smartwatch monstrosity and make it look purdy. Feel free to get creative with the casing. You can use whatever materials you have available. What I did was take the measurements for the watch, the screen, the backlight button, the on/off switch, and the micro USB charger and enter them into a CAD program. For simplicity's sake, I used Tinkercad[19], an online CAD program. The watch size ended up being about 1 ¾ inches wide and tall, which is tolerable. But it also ended up being one inch thick, which is not as desirable. Most of that space was being used up by the battery I

19 https://www.tinkercad.com/

chose. So in retrospect, I'll probably end up going with a thinner but wider battery.

I took the CAD model and printed it out using a 3D printer. It took a couple of tries before I was able to get everything to fit. To finish things off, I added the screen protector that I salvaged from the original cell phone along with a watch band. Overall, I'm happy with the results. If I were to revisit this project, I'd probably opt for a slimmer battery, smaller Bluetooth, and a fabricated PCB. Hopefully this inspires some of you to upcycle your own old cell phones into something cool!

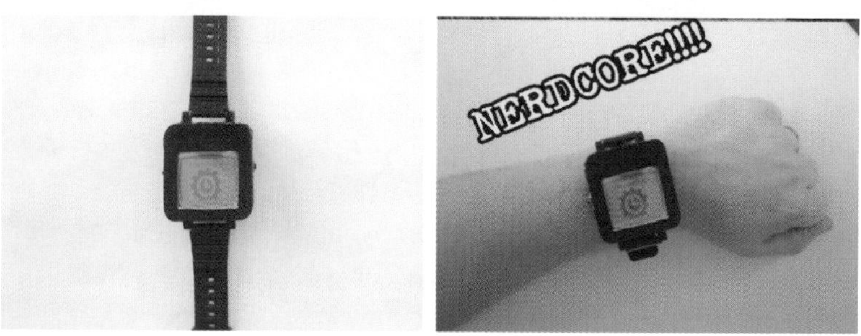

dissentiet vim.

Invidunt neglegentur cu per, labores gubergren voluptaria id duo. An choro maiestatis consequuntur pri. Homero munere nominavi eum no, an sit idque doctus tincidunt. Saepe legimus euripidis cum ut. Dicit tation scripserit vim no, illum commune scriptorem nec in. No vis dicam partem.

eu. I
ri, ea
petua pericula
Isu ut
endo
n hendrerit

voluptaria
omero munere
Saepe
it vim no,
artem.

PHOTO GLOSSARY

The following photo glossary is here to help out with the detail on the projects. You'll find enlarged images of some of the projects in the next few pages in an attempt to make them as clear as possible.

Enjoy the upcycling!

Difficulty

OLD WEBCAM TO BACKUP CAMERA

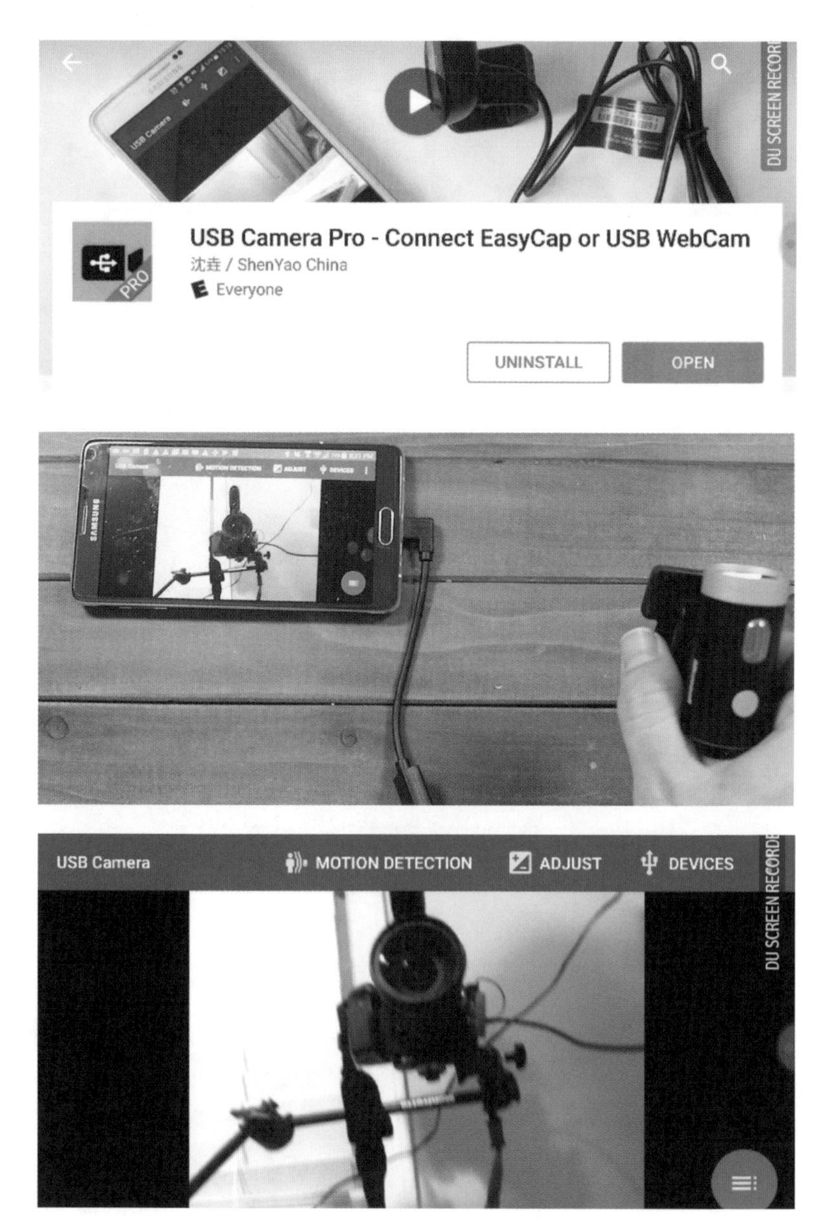

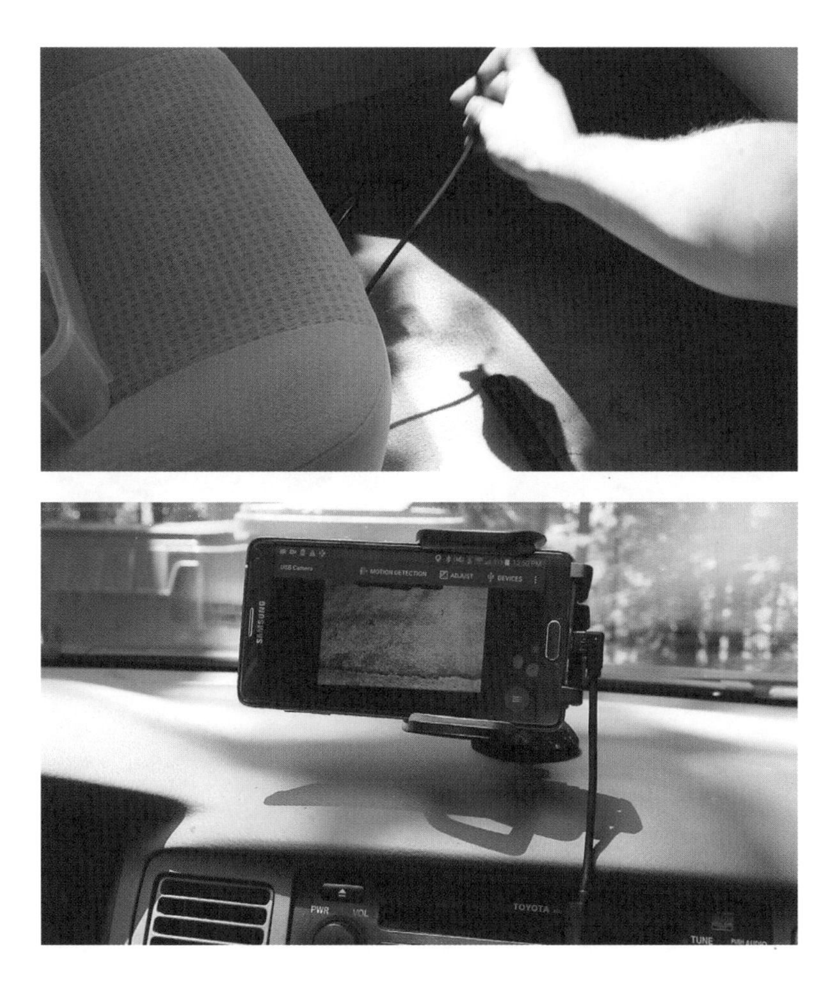
USB Camera
MOTION DETECTION
ADJUST
DEVICES
SAMSUNG
PWR
VOL
TOYOTA
TUNE

Screen
Bezel

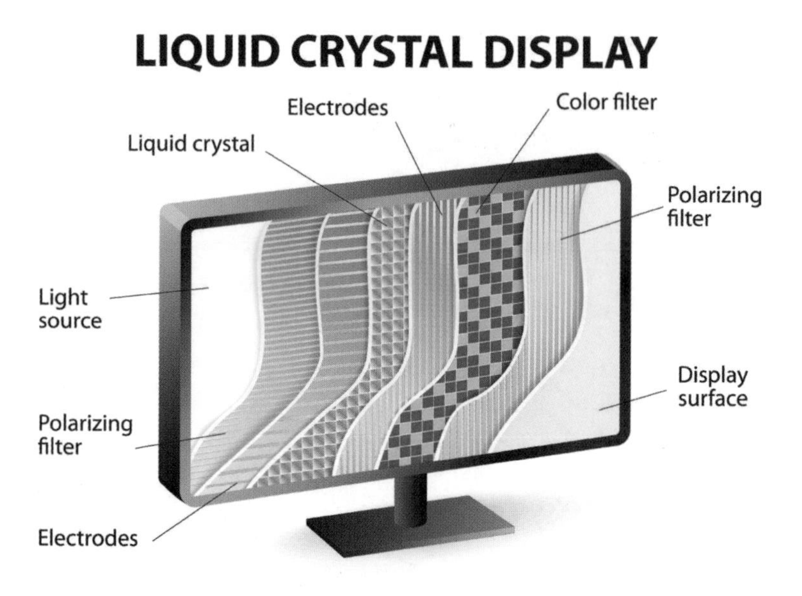
LIQUID CRYSTAL DISPLAY
Electrodes
Color filter
Liquid crystal
Polarizing filter
Light source
Display surface
Polarizing filter
Electrodes

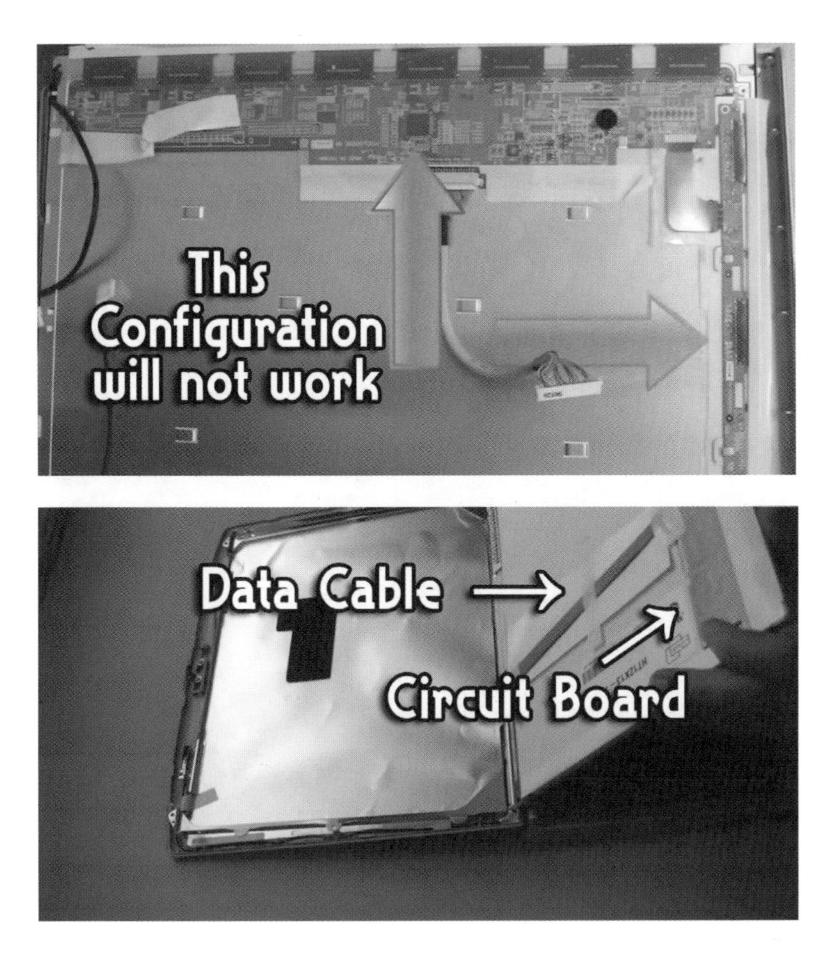
This
Configuration
will not work
Data Cable
Circuit Board

CD-ROM DRIVE TO 3D PRINTER

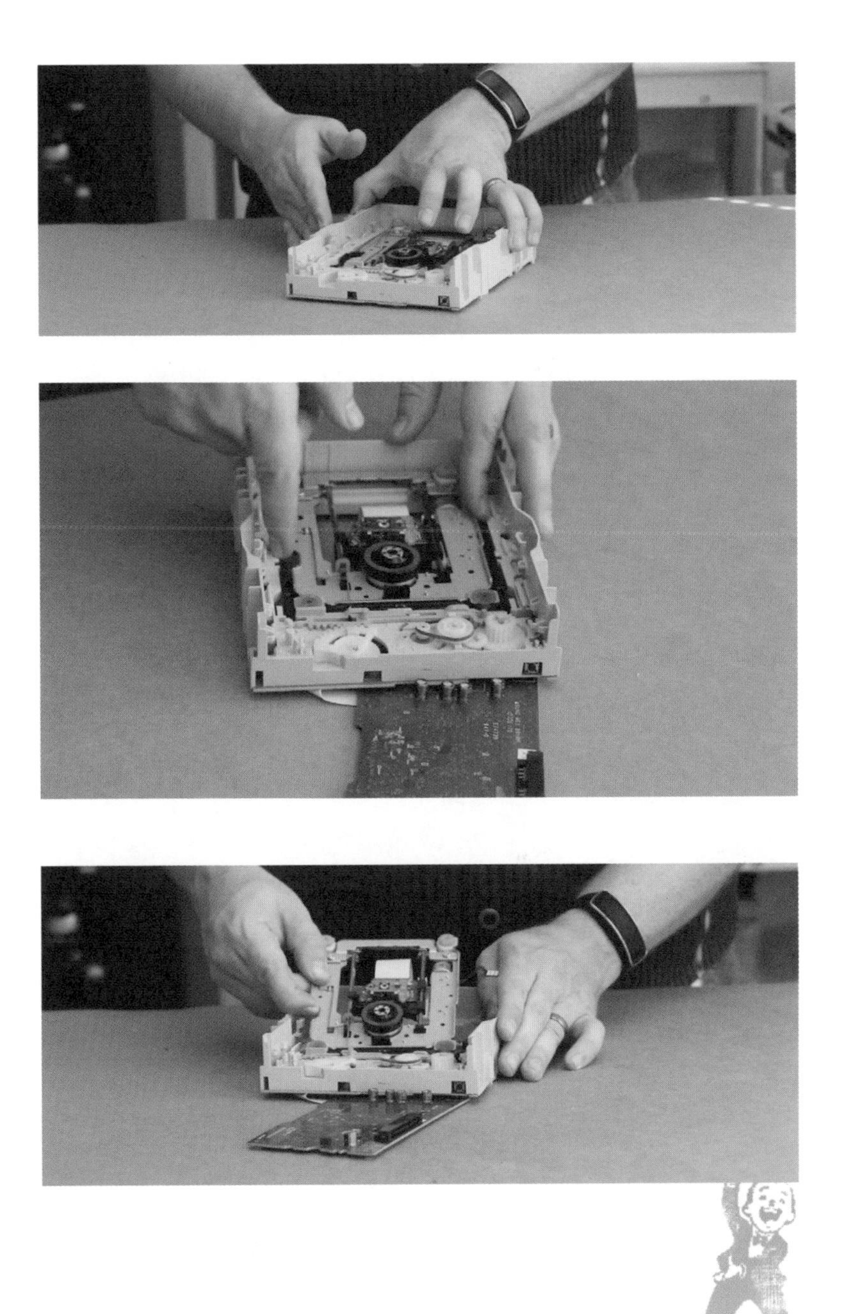

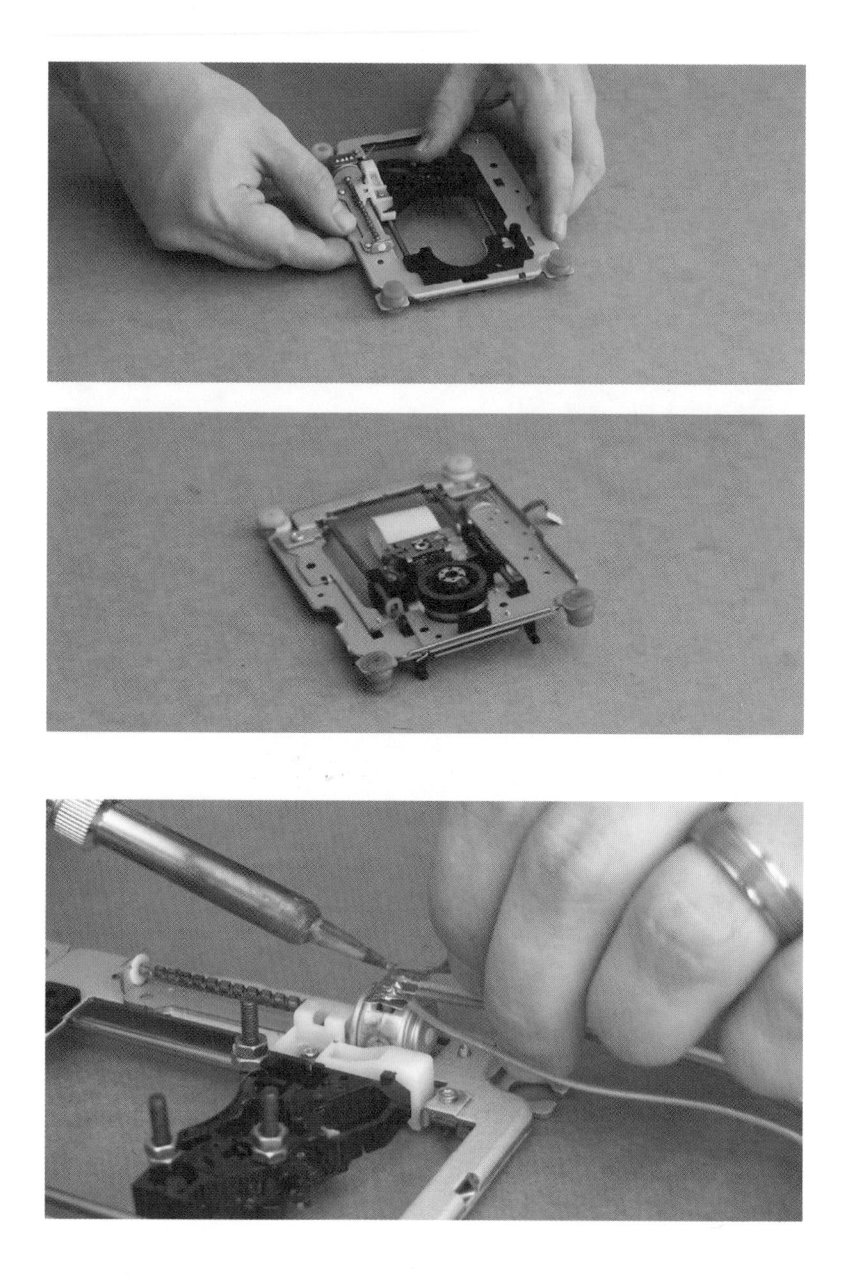

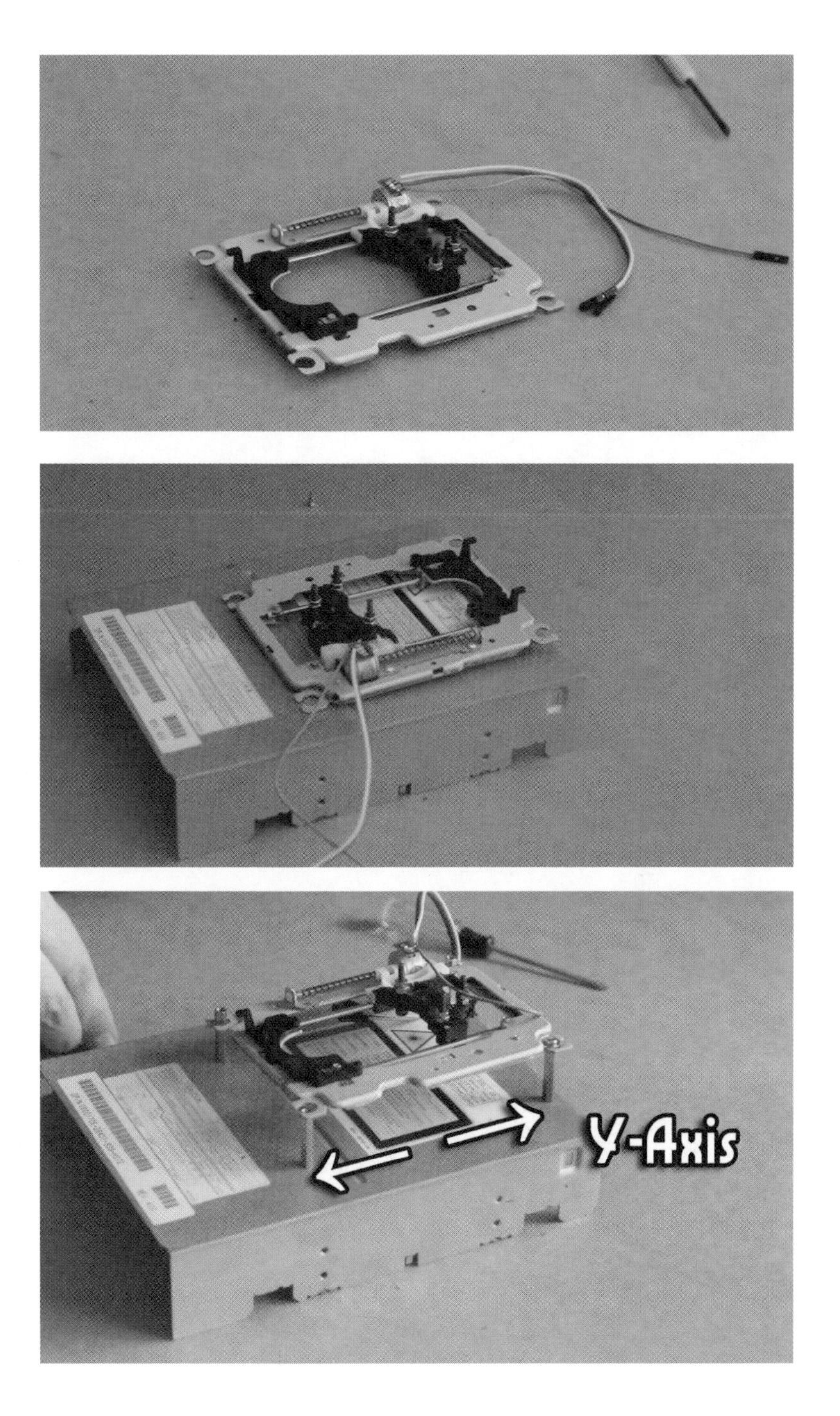
Y-Axis

X-Axis
Z-Axis

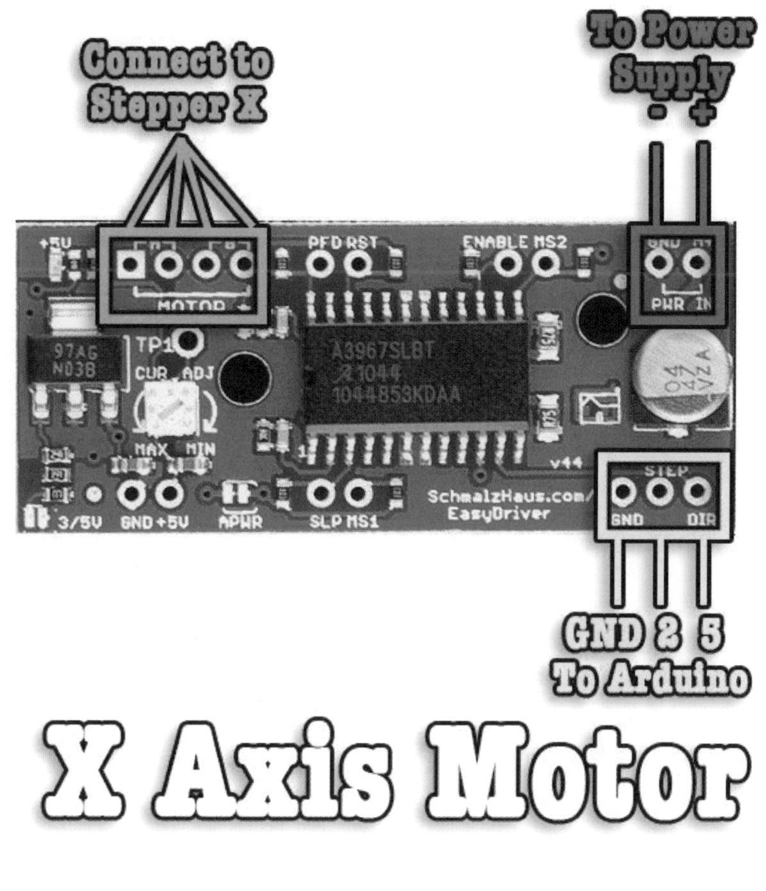
Connect to Stepper X
To Power Supply
- +
PFD RST
ENABLE MS2
PWR IN
97AG
N03B
TP1
CUR ADJ
A3967SLBT
1044853KDAA
MAX MIN
v44
STEP
3/5V GND +5V
APWR
SLP MS1
SchmalzHaus.com/
EasyDriver
GND DIR
GND 2 5
To Arduino

X Axis Motor

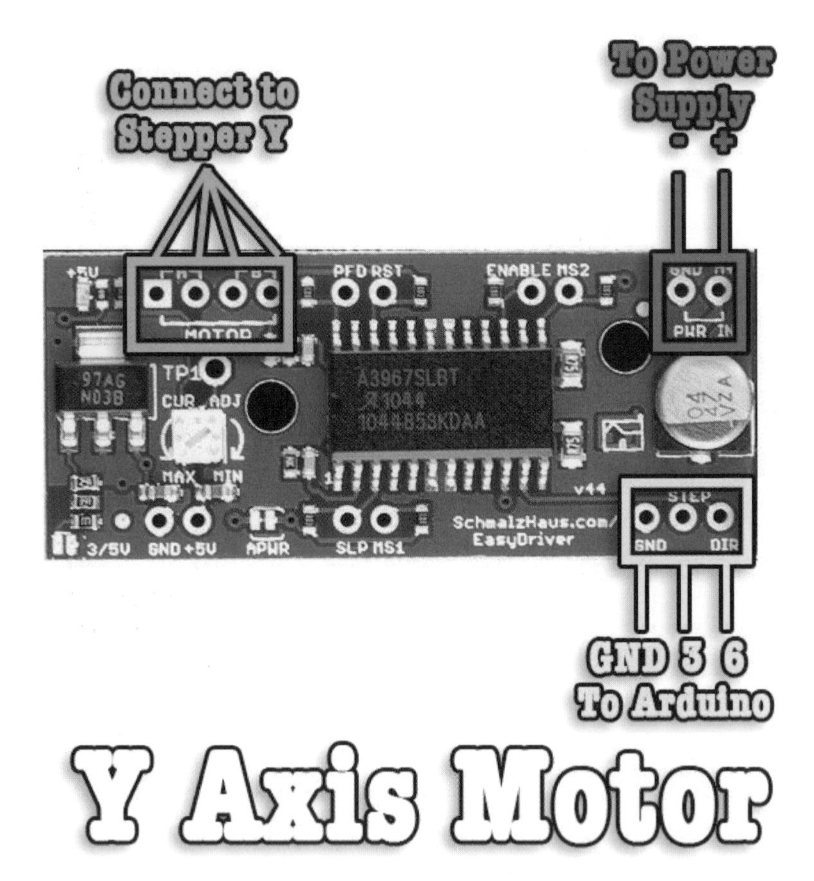
Connect to Stepper Y
To Power Supply
- +
GND 3 6
To Arduino
Y Axis Motor

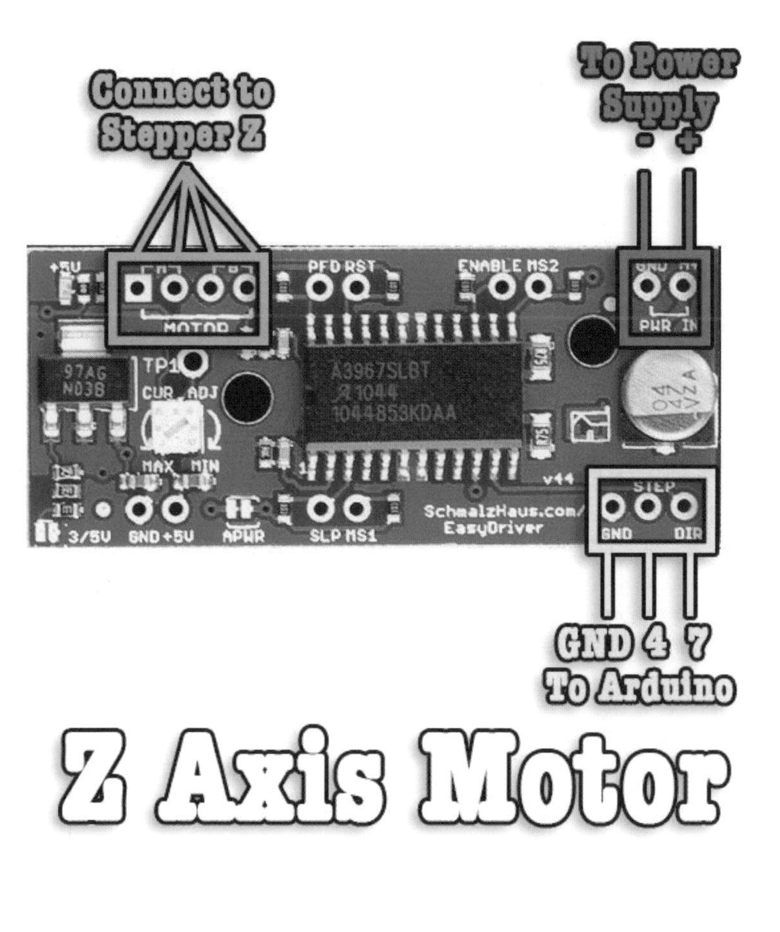
Connect to Stepper Z
To Power Supply
- +
+5V
MOTOR
PFD RST
ENABLE MS2
PWR IN
TP1
CUR ADJ
A3967SLBT
1044853KDAA
MAX MIN
v44
STEP
GND DIR
3/5V GND +5V
APWR
SLP MS1
SchmalzHaus.com/
EasyDriver
GND 4 7
To Arduino

Z Axis Motor

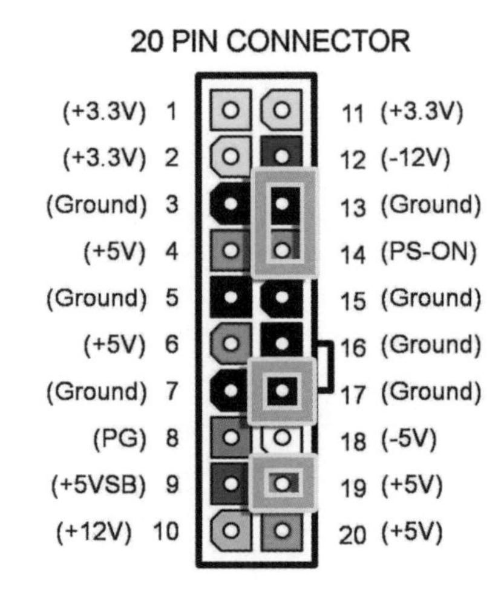
20 PIN CONNECTOR
(+3.3V) 1
(+3.3V) 2
(Ground) 3
(+5V) 4
(Ground) 5
(+5V) 6
(Ground) 7
(PG) 8
(+5VSB) 9
(+12V) 10
11 (+3.3V)
12 (-12V)
13 (Ground)
14 (PS-ON)
15 (Ground)
16 (Ground)
17 (Ground)
18 (-5V)
19 (+5V)
20 (+5V)

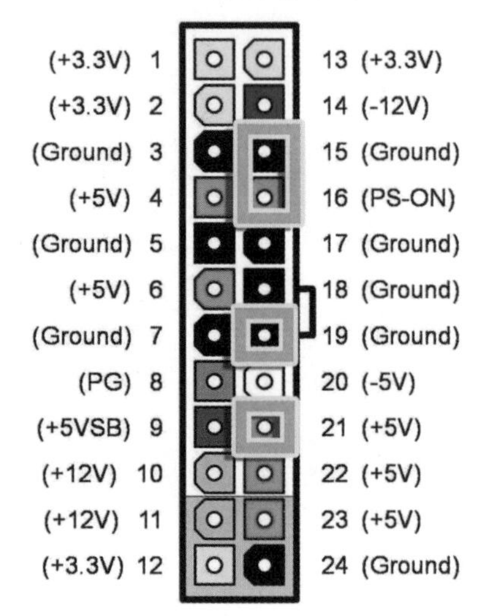
24 PIN CONNECTOR
(+3.3V) 1
(+3.3V) 2
(Ground) 3
(+5V) 4
(Ground) 5
(+5V) 6
(Ground) 7
(PG) 8
(+5VSB) 9
(+12V) 10
(+12V) 11
(+3.3V) 12
13 (+3.3V)
14 (-12V)
15 (Ground)
16 (PS-ON)
17 (Ground)
18 (Ground)
19 (Ground)
20 (-5V)
21 (+5V)
22 (+5V)
23 (+5V)
24 (Ground)

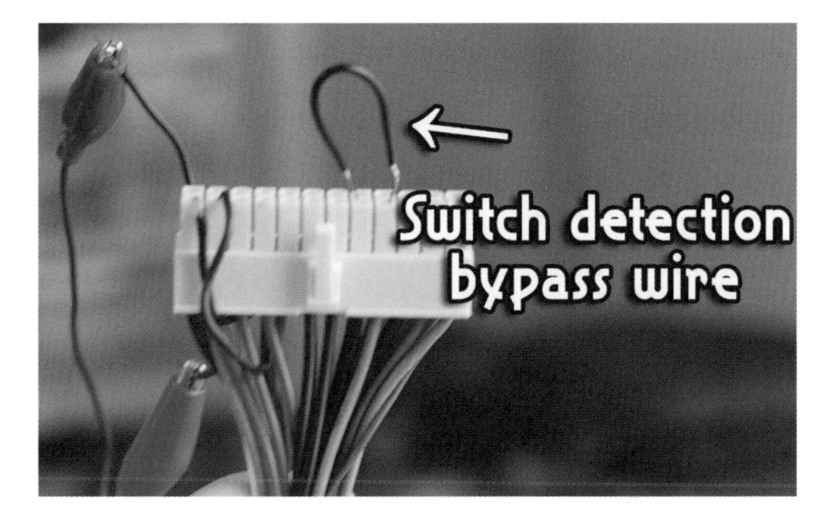
Switch detection
bypass wire

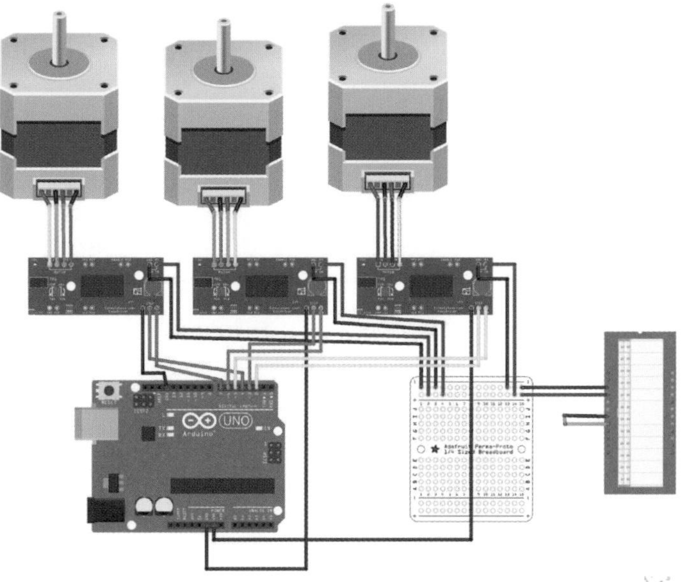

Grbl Controller 3.5.1

File Tools Help

Port name COM7

Close / Reset

Baud Rate 9600

Last State: Idle

Send File

Choose file

Begin Stop

File progress 0%

Queued Commands 0

Runtime:

Command

```
$2=250.000 (z, step/mm)
$3=10 (step pulse, usec)
$4=250.000 (default feed, mm/min)
$5=500.000 (default seek, mm/min)
$6=192 (step port invert mask, int:11000000)
$7=25 (step idle delay, msec)
$8=10.000 (acceleration, mm/sec^2)
$9=0.050 (junction deviation, mm)
$10=0.100 (arc, mm/segment)
$11=25 (n-arc correcti
on, int)
$12=3 (n-decimals, int)
$13=0 (report inches, bool)
$14=1 (auto start, bool)
$15=0 (invert step enable, bool)
$16=0 (hard limits, bool)
$17=0 (homing cycle, bool)
$18=0 (homing dir invert mask, int:00000000)
$19=25.000 (homing feed, mm/min)
$20=250.000 (homing seek, mm/min)
$21=100 (homing debounce, msec)
$22=1.000 (homing pull-off, mm)
ok
Spindle On.
> M03
```

Machine Coordinates

X

Y

Z

Axis Control Visualize

Absolute coordinates

Spindle On

Zero Position Go Hom

Machine Coordinates (mm) Work Coordinates (mm)

X 0.000 0.000

Y 0.000 0.000

Z 0.000 0.000

Axis Control | Visualizer | Advanced

Z Jog

0

0

Absolute coordinates after adjust

Spindle On

Step Size

10

Zero Position | Go Home

OLD SMARTPHONE TO SECURITY CAMERA

Webcam
Not secure | 192.168.0.13:8080
Webcam Home Video archive Videochat drivers Other viewing methods

Video renderer: No video | Flash | Browser | Java | Javascript | Fullscreen
Audio player: No audio | Flash | HTML5 Wav | HTML5 Opus | Why the lag?
Two-way audio

order control
Enter a label for this recording
ular recording chunk length
ircular recording records video in chunks of 1 hour, overwriting
e older chunks when storage space is running out.
tos
Take photo | Take focused photo
Save photo to storage | Save focused photo to storage

Zoom
Stream quality
Exposure compensation
Misc: Autofocus hold | LED Flashlight | Overlay | Night vision
Front camera

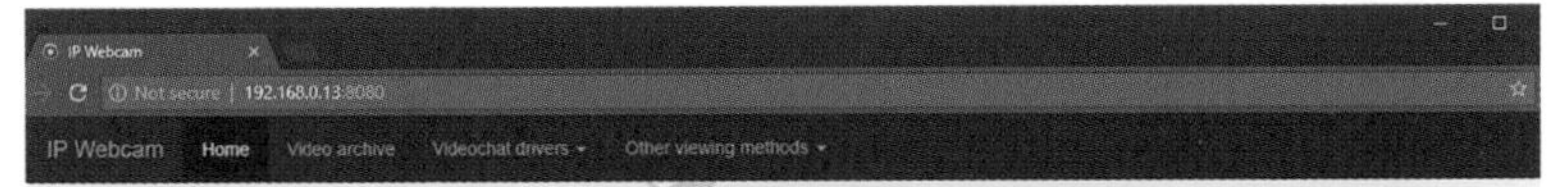

IP Webcam

Not secure | 192.168.0.13:8080

IP Webcam Home Video archive Videochat drivers Other viewing methods

Recorder control

Enter a label for this recording

Circular recording chunk length

Circular recording records video in chunks of 1 hour, overwriting the older chunks when storage space is running out.

Photos

Take photo | Take focused photo

Save photo to storage | Save focused photo to storage

Tasker events control What is this?

Zoom 1 X

Stream quality 49%

Exposure compensation 0

Misc Autofocus hold | LED Flashlight | Overlay | Night vision

Front camera

Motion detection Enabled | View areas

More sensitivity **Less sensitivity**

Open sensor graph »

Motion detection areas

Advanced settings

REVIVE AN OLD IPOD

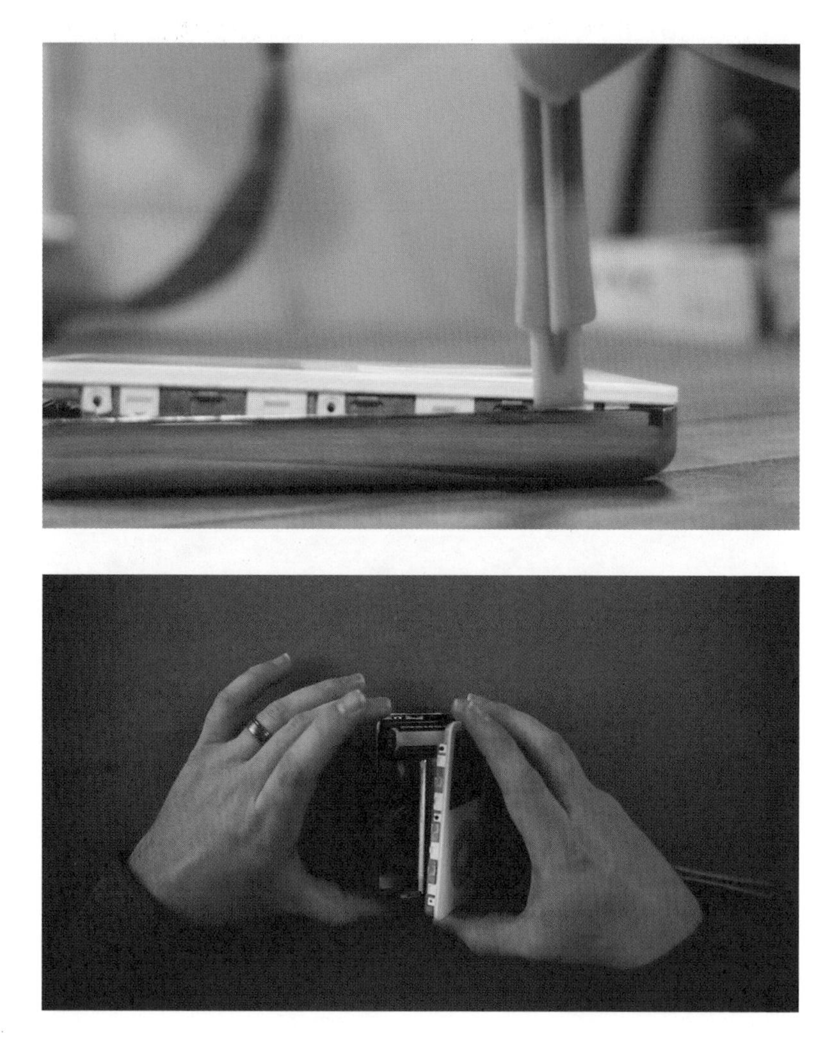

32GB

32GB

OLD CELL PHONE TO SMARTWATCH

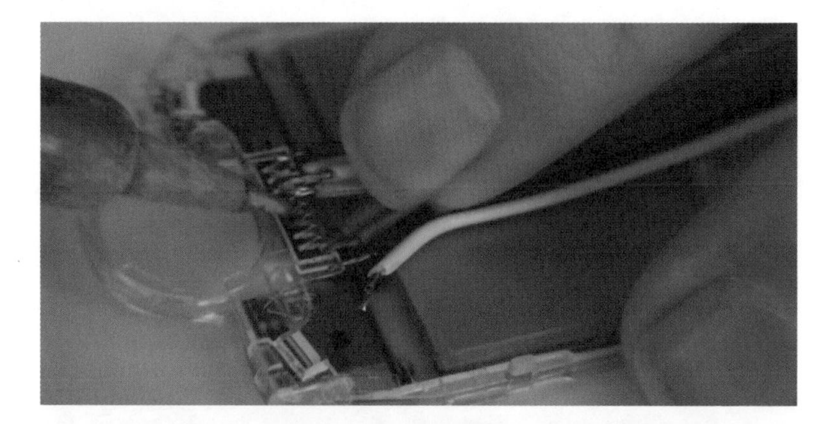

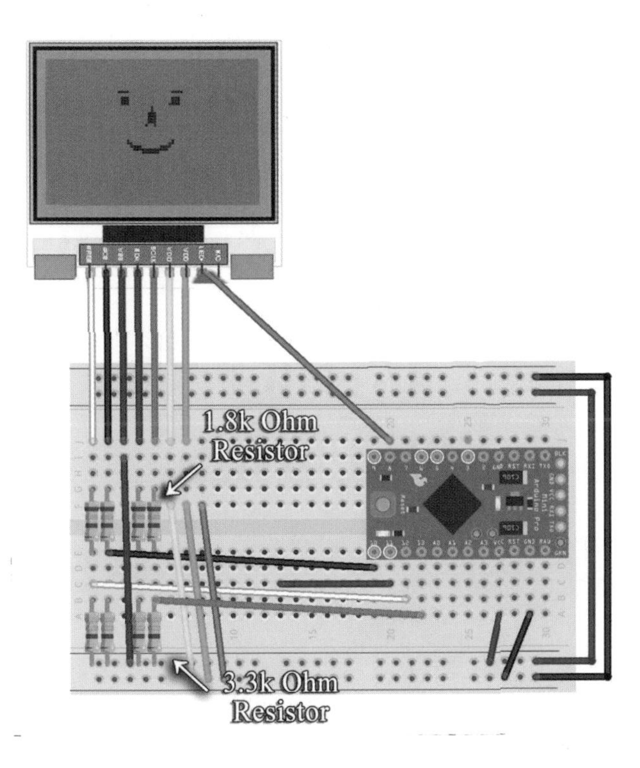
1.8k Ohm Resistor
3.3k Ohm Resistor

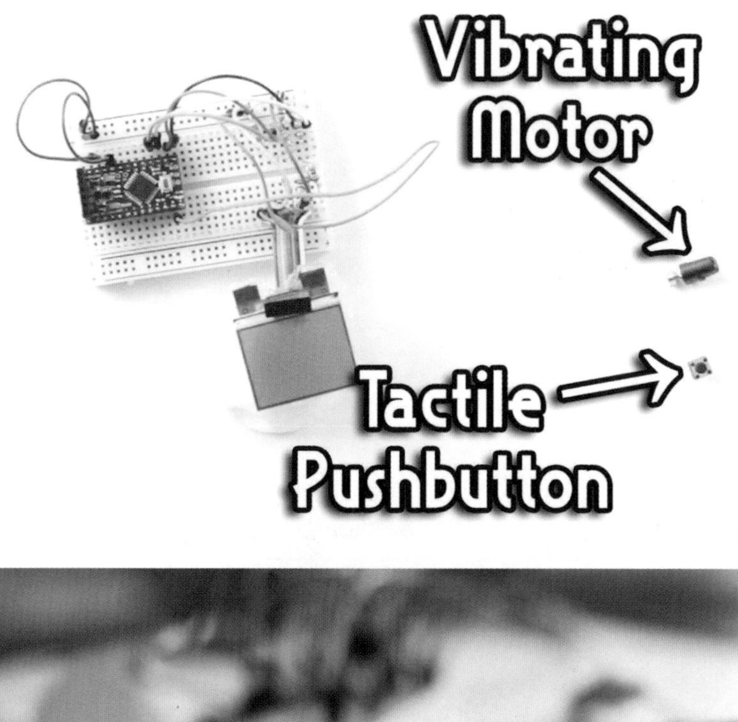
Vibrating Motor
Tactile Pushbutton

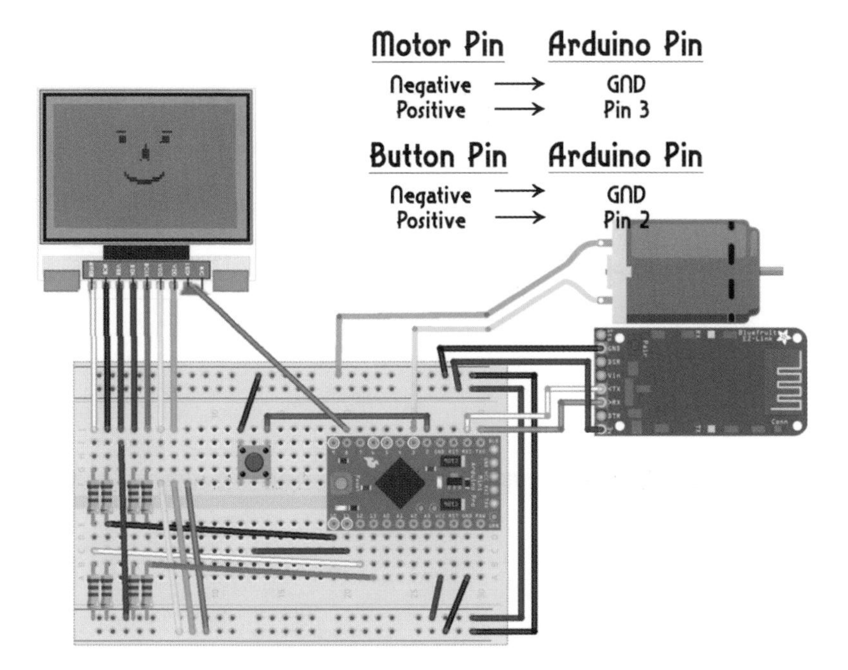
Motor Pin
Arduino Pin
Negative → GND
Positive → Pin 3
Button Pin
Arduino Pin
Negative → GND
Positive → Pin 2

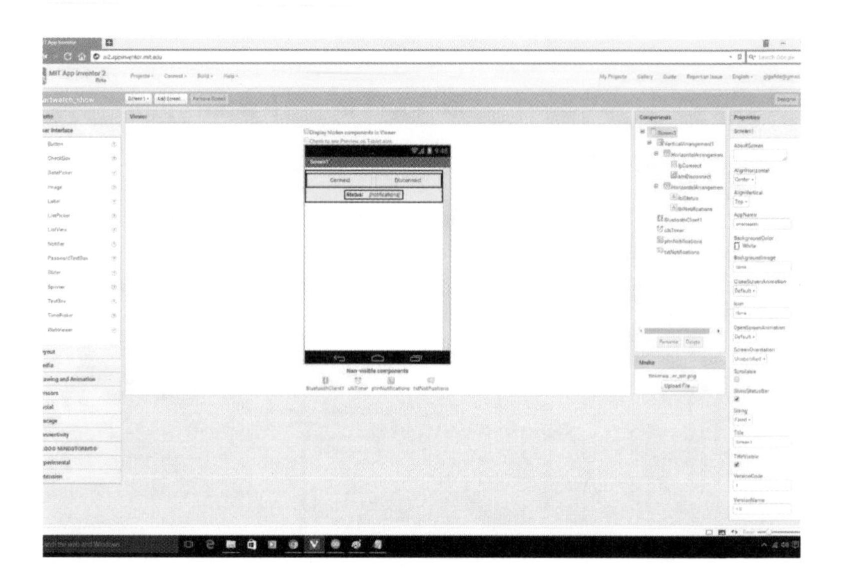

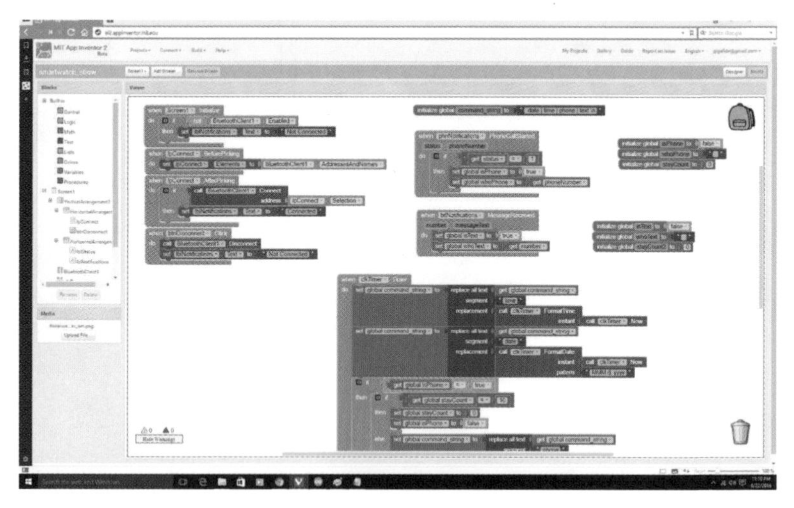

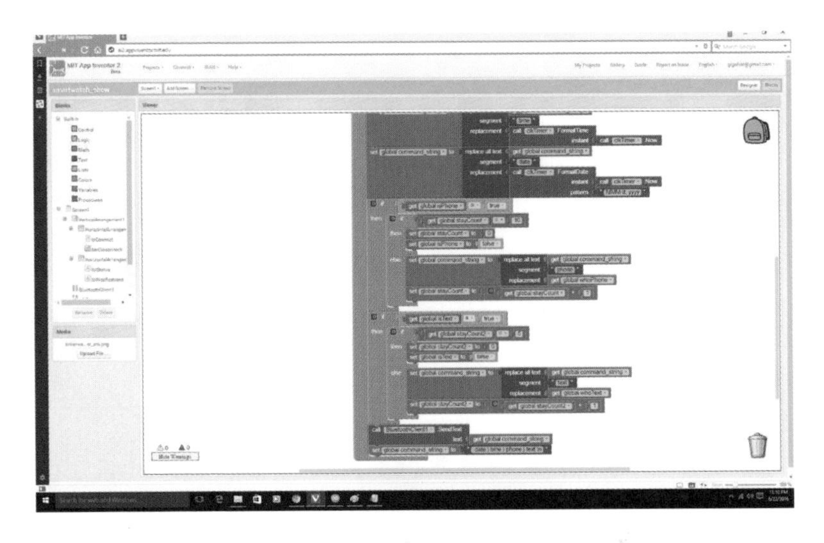

Tinker Watch

Connect

Disconnect

Status: *[notifications]*

Steps: (not tracking)

Notification Options:

Enable Phone Notifications

Enable Text Notifications

Enable Pedometer

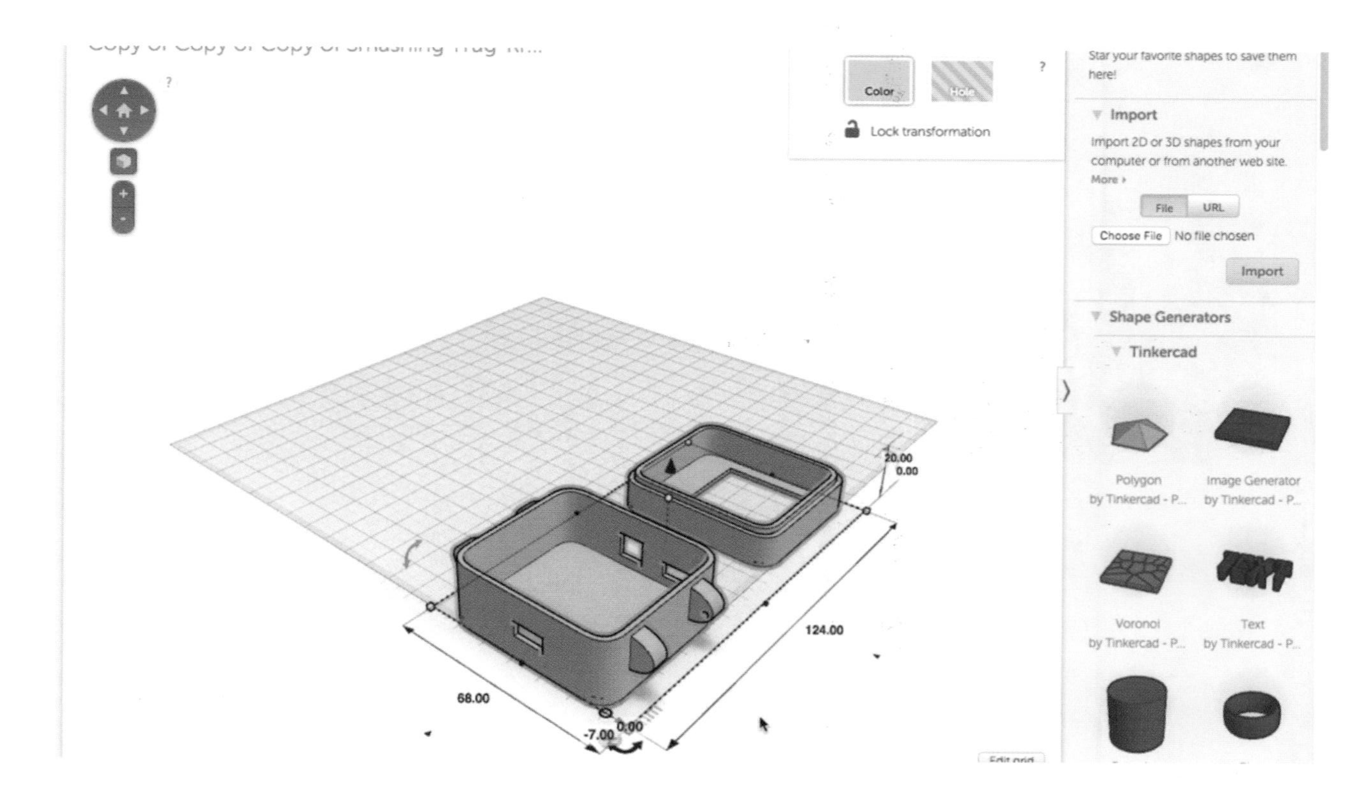
Color
Hole
Lock transformation
Star your favorite shapes to save them here!
Import
Import 2D or 3D shapes from your computer or from another web site.
More
File
URL
Choose File
No file chosen
Import
Shape Generators
Tinkercad
Polygon
by Tinkercad - P...
Image Generator
by Tinkercad - P...
Voronoi
by Tinkercad - P...
Text
by Tinkercad - P...
20.00
0.00
124.00
68.00
-7.00
0.00

dissentiet vim.

Invidunt neglegentur eu per, lobortis gubergren voluptaria id duo. An choro maiestatis consequuntur pri. Homero munere nominavi eum no, an sit idque doctus tincidunt. Saepe legimus euripidis cum ut. Dicit tation scripserit vim no, illum commune scriptorem nec in. No vis dicam partem.

u. I
i, ea
etua pericula
eu ut
ndo
hendrerit

voluptaria
mero munere
Saepe
t vim no,
artem.

PROJECT JOURNAL

Each project is unique to its complexities. Therefore, the best way to approach them is to keep track of the details. This journal here allows you to check-off your tools and items, measurements and steps and anything else you need as you embark on the upcycling challenge.

Difficulty

OLD WEBCAM TO BACKUP CAMERA

Parts & Tools Check List

- USB webcam
- Android smartphone
- Six foot USB extension cable
- USB OTG adapter
- Strong magnets

TURNING AN OLD LAPTOP INTO A PROJECTOR

Parts & Tools Check List

- Old unused laptop
- Overhead projector
- Screwdriver

CD-ROM DRIVE TO 3D PRINTER

Parts & Tools Check List

- 3 x Desktop computer CD-ROM or DVD-ROM drives
- 3 x Stepper motor drivers
- 1 x Arduino
- 1 x Desktop computer power supply
- 2 x Electrical box covers
- 1 x Generic 3D printing pen
- 1 x 22 Ohm resistor (value may vary depending on your 3D printing pen)
- 1 x NPN transistor
- Various nuts, bolts, and spacers
- Soldering equipment and wire

OLD SMARTPHONE TO SECURITY CAMERA

Parts & Tools Check List

- Old Android smartphone

REVIVE AN OLD IPOD

Parts & Tools Check List

- An old iPod Classic (may not work with other models)
- A CompactFlash (CF) camera card
- CF Card to ZIF adapter (available on eBay, Amazon, or Ali Express)
- iPod Classic repair kit (also available on eBay, Amazon, or Ali Express)

OLD CELL PHONE TO SMARTWATCH

Parts & Tools Check List

- Old Nokia Cell Phone
- SparkFun Arduino Pro Mini 3.3v
- Bluetooth Module
- Micro USB charging board
- 3.7v Lithium ION battery
- 4 x 1.8 kOhm resistors
- 4 x 3.3 kOhm resistors
- 1 x 10 kOhm resistors
- 1 x diode
- Small slide switch
- Small momentary button

t vim.

[illegible]
a choro maiestatis consequuntur pri. Homero munere
eum no, an ait idque doctus tincidunt. Saepe
euripidis cum ut. Dicit tation scripserit vim no,
mune scriptorem nec in. No vis dicam partem.

AFTERWORD

Upcycling is more than just slapping a new coat of paint on a night stand. There's an element of artistry and nostalgia in old things that we often take for granted in the wake of silicon chips and mass production. Each item has a history and tells a story of a bygone era and how we've developed as a society. To let time and lack of interest steal these hidden treasures from us is like leaving a music box's melody to be buried and forever lost in a landfill.

Through upcycling, we are not only keeping these legacies alive; we are adding to them. Just like adding a dash of spice to an old recipe, we are creating art while building upon the art of others. Each item that we upcycle has an endless history of artistry and innovation: how the materials were sourced, the tools that made it, the design process, the production process, and the path it took to get to you. It's the lineage of our society, and it's that lineage that we are honoring and continuing.

As in life, there can be tremendous benefit in taking a second look at what's considered worthless junk and embracing it instead of discarding it. Inspiration can be found in things you would never expect. Hopefully this book has taught you to see through the label of "junk" and to instead focus on the potential of older gadgets to become something new. Maybe one day, another generation will upcycle your upcycled creations!

ACKNOWLEDGEMENTS

An acknowledgement for my own book. As an IT professional, that's not something I ever thought I'd be writing. Don't get me wrong, I've often wondered what it would be like to write a book and an acknowledgement, but I never really expected that wonder to become an actual, legitimate reality. This whole process of writing a book has been surreal and somewhat eye-opening. Being a part of this process and seeing "how the sausage is made," as it were, has really brought to light the hidden roles of the people that surround me. So even though this is slightly unfamiliar territory for me, the only reason this book even exists is because of the love, kindness, guidance, cooperation, patience, and encouragement of others, and I want to do my best to acknowledge those hidden figures.

Putting a book together requires a lot of time and attention. But when you have a family, including a one-year-old, having spare time and attention is a precious commodity. To that end, I have to thank my wife, Marian, who sacrificed her own time and attention and spent many a night with a screaming toddler just so that I could see this thing through. For me, love is shown more by actions than words, and to cede her time so that I could achieve my goals speaks more than a book full of love notes.

The topics and projects in this book are the result of my passion for learning and understanding technology. I share this passion with the thousands of viewers and followers on my YouTube channel. Sometimes it's hard to continue having a presence on social media due to all the negativity surrounding those platforms, but I have a great base of subscribers and commenters on my channel who have spurred me on and supported me for years. If it wasn't for them, I would have given up on making stuff a long time ago.

How you grow up shapes who you become when you're older. You always hear that growing up in a large family is great, and I had the

delightful benefit of being raised in a large family with my mom and dad, two brothers, and three sisters. All of them have always encouraged and supported my drive and creativity...at least to my face. Aside from my family, my biggest influence growing up was Lewis H. Hunt, our 70-year-old cattle-farming neighbor. He showed me the value of hard work and how to push through the boundaries of my perceived limitations. He was an amazing guy, and he will be missed.

Let's be honest, if I were to try and publish a book on my own, it would probably make for the world's greatest toilet paper. All I have is a jumbled mess of thoughts. What you're holding in your hands right now is the product of a team of hard workers who are experts at turning monochromatic words into art. I am in humble gratitude to the team at Mango for giving me this once-in-a-lifetime opportunity!

Finally, I would like to thank my friends at SplatSpace, my local hacker/maker space. Maker spaces in general are a great place to meet likeminded makers, share ideas, and get feedback for projects. SplatSpace is a great group of men and women who love to look at the world and see what makes it tick. They've inspired me and helped me refine myself as a maker. Thank you all!

ABOUT THE AUTHOR

Daniel Davis is an IT professional and Youtube personality with an insatiable curiosity and a propensity for tinkering. Using his desire to constantly learn and share what he's learned, Daniel founded the STEM focused Tinkernut, LLC (www.tinkernut.com) to teach people more about tinkering, hacking, and programming. His efforts through Tinkernut, LLC have been fetured and published in several major websites, blogs, textbooks, and magazines. Aside from learning and teaching, Daniel is also an active supporter and member of local maker spaces and maker space initiatives.